Arthur Lakes

Prospecting for Gold and Silver

Arthur Lakes

Prospecting for Gold and Silver

ISBN/EAN: 9783337077204

Printed in Europe, USA, Canada, Australia, Japan

Cover: Foto ©berggeist007 / pixelio.de

More available books at **www.hansebooks.com**

PROSPECTING

FOR

GOLD AND SILVER.

BY

ARTHUR LAKES,

Late Professor of Geology at the State School of Mines,
Golden City, Colorado.

Author of "GEOLOGY OF COLORADO AND WESTERN ORE DEPOSITS,"
"GEOLOGY OF COLORADO COAL DEPOSITS," Etc.

SCRANTON, PA.
THE COLLIERY ENGINEER CO

1895.

PREFACE.

In preparing this little work the author has felt the difficulty which arises in a theoretical dissertation on so eminently practical a subject as prospecting. It seems like giving rules and prescriptions for hunting or fishing or any other natural or practical pursuit. Though theory and practice are not at variance when happily combined, yet either without the other proves very unsatisfactory. Thus the reader of this book, should he start out armed only with its theory, will find himself for some time pretty much "at sea" when he comes to actual practice in the field. As, however, he gradually obtains some practical experience, he may find this little work of use to him. So, also, the seasoned prospector, who has hitherto trusted to luck, keenness of observation, intuition and experience, may find himself in the future much better equipped by acquiring a little of the theory.

Whilst we have endeavored to give the prospector all assistance in our power, as to the best means of educating himself, describing his outfit, etc., we have devoted special attention to the description of such geological and other phenomena as he is likely to meet with in connection with his work, so that he may have an intelligent idea of them when he encounters them.

We have selected just as much material as we think would be most interesting and useful to him, saving him the time and trouble of wading through heavy tomes and laboriously picking out from a vast amount of, for his purpose, superfluous matter, that which he will most require.

The work is intended to be a popular one, addressed to the average student, prospector and miner and to the general public. The illustrations are largely drawn by the author from Colorado mines and Rocky Mountain subjects which are most familiar to him.

ARTHUR LAKES,
State School of Mines,
Golden City, Colorado.

JANUARY 1, 1895.

CONTENTS.

PROSPECTING FOR GOLD AND SILVER.

CHAPTER I.

ON PROSPECTING—PREPARATION AND OUTFIT FOR WORK.

The regular prospector, as a rule, has at some time of his checquered career had some actual experience in the mines themselves, from which he has learned by observation, the appearance of different ores, their different values, how the veins appear on the surface, how to open a vein, and the uses of pick, shovel, and blasting powder. In a word he is a miner, who has become too restless to stick to steady work, and so follows the more uncertain and precarious livelihood of seeking for new and undiscovered veins, many of which even in an old mining district may yet be discovered covered up by brush or debris, whilst a new district offers a most enticing field. These mineral veins or ledges, may make him in a moment a comparatively rich man, and if he finds them, they will cost him nothing, only a simple compliance with the inexpensive regulations of the law. So the life of a prospector offers many attractions to one who is restless and loves to roam and loves to find something new and is not afraid of considerable hardship. To save a vast amount of time and labor, he should acquire knowledge. Thus, for instance, if he were prospecting for coal he would be wasting his time in hunting for it in granite, or if he was hunting for the precious metals, he would lose time in looking for them among the unaltered sedimentary strata of the prairie. This is merely for example, but an infinite variety of knowledge is necessary for him in his vocation, besides even that of the simpler elements of geology, such as the knowledge of different kinds of minerals, and their value, the kind of places and peculiar rocks they are associated with, their appearance on the surface, etc., etc.

together with some knowledge of assaying or blowpiping or panning.

In a newly discovered camp, men will rush in for a few weeks, work a little in the different mines, sufficient to give them an idea of the kind of ores and rocks and other circumstances in the locality, and then will strike out on their own account and prospect around the camp for new veins or extensions of those already discovered. An extension, by the way, of a very rich discovered lode is not always to be relied on. Nature seems often to concentrate her riches at one point, and leave the extension barren, as in the case of the Comstock of Nevada. But little wealth has been found outside of the great lode and mine itself.

The best education is in the mines themselves, so a novice on arriving at a mining region had better spend as much time as possible in practical work, in, and around the various mines, before he launches out prospecting. A prospector can rarely carry about much assaying or other apparatus with him for determining the character or value of ores he may find, and hence it is well for him to accustom himself to these ores in the mines themselves. Also he should acquaint himself with the peculiar ores of each particular district, before he attempts to prospect in its vicinity, for an ore such as coarse grained galena, in one district may be generally rich, whilst in another it is remarkably poor in silver.

The best previous education for a prospector would be a course at a school of mines, where he will learn the elements of geology, mineralogy, assaying, etc. And next to that, practical work in the mines themselves, and lastly the prospecting field. A little knowledge of blowpiping may also help him, which he may acquire at his school.

Having left his school, he should learn the practical use of the pick, drill, and, blasting powder. By working around a concentrator he will learn the difference between ore and gangue rock ; and "picking" or "sorting" ores, will teach him at sight the values of ores. The prospector should know how to open his vein or ledge, when he finds it, with pick, shovel, and blasting apparatus. A little carpentry will teach him how to make a handwinch, and a few lessons in blacksmithing, will teach him how to sharpen and temper his tools, for there will probably be no blacksmith's shop or carpenter's either, within miles of where he may go. Other prospectors will teach him how to use his pan or iron spoon for testing ores, and various other dodges and make-shifts. An important point is to learn how to average approximately the quantity of ore in, and value of, a ledge when he has found one. Valuable ore on a ledge lies in pockets, strings, bunches, irregularly distributed through the quartz or other material of the vein; he should learn to tell at sight the relative proportion of ore and gangue. He would do well to study the result of working ores in a mill or furnace, such as trying to estimate the yield of bullion of the ores which are mined, taking them in weekly or monthly lots. With some such preliminary knowledge he is ready for the field.

HIS OUTFIT.

The following list of necessaries by Mr. A. Balch in his "Treatise on Mining" is as full as can be given by any one, and is more than the average prospector generally needs.

A PROSPECTOR AND HIS OUTFIT.

"*First.* Two pairs of heavy blankets weighing about 8 pounds each.

Second. A buffalo robe or a blanket lined poncho.

Third. Suit of strong gray woolen clothes, pair of brown jean trousers, a change of woolen underclothing, woolen socks, pair of heavy boots, soft felt hat, three or four large colored handkerchiefs, a pair of buckskin gauntlets, toilet articles, etc. All should go into a strong canvas bag.

Fourth. A breech loading rifle or shot gun and a revolver. Around his waist a strong sash to carry his holster and knife, in a sheath. His ammunition, if his revolver is large bore, may conveniently fit both his rifle and revolver. Pipe and tobacco.

Fifth. A sure footed native or mountain pony. A Mexican saddle with its saddle horn, straps, etc., to tie on various things, such as his pack, bags, water canteen, etc. The left stirrup may be fitted with a leather tube, in which the rifle barrel may be placed. A strap around the saddle horn will secure the gun stock. The long lariat or stake

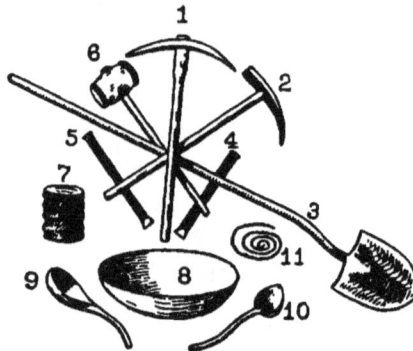

A PROSPECTOR'S TOOLS.

1, 2. Picks.
3. Long handled Shovel.
4, 5. Drills.
6. Heavy Hammer.
7. Blasting Powder.
8. Pan.
9. Horn Spoon.
10. Iron Spoon.
11. Fuse.

rope for tethering his horse should be coiled up and tied by a strap to the saddle horn.

Sixth. For prospecting, a 'poll' pick and prospecting pan made of iron or a horn spoon should be carried. The pan is also useful besides for washing out sand, as a dish or bathing vessel. A large iron spoon for melting certain metals is likewise to be carried, and in some cases a small portable Battersea assaying furnace.

Seventh. A frying pan 8 inches diameter of wrought iron, a coffee pot, tin cup, spoon, and fork, and matches in tin box, pocket compass, a spy glass, or pair of field glasses.

Eighth. Provisions, bacon, flour, beans, coffee, or tea, pepper, salt, and box of yeast powder, all packed in strong bags, to go into a canvas sack. A few lessons in the kitchen on cooking will be advantageous before starting.

Ninth. Packing the bronco. Place a folded blanket on the horse's back, on this lay the saddle. The saddle bags contain small things. The bags with provisions are placed behind the cantle of the saddle; on top of this the bag of clothing. The pick goes on top tied by a thong. Coffee pot, and frying pan are lashed on the bags."

Sometimes a prospector takes a horse to ride on and another as a pack animal, or a donkey only. For grass and water for his horse, he must trust to the country. He will fix his temporary camp in some suitable location, where these are to be found, and thence, as from headquarters, prospect daily the adjacent country returning nightly, it may be, to his camp.

BRIEF SKETCH OF PROSPECTING.

We may divide the prospecting for the precious metals into two general classes: hunting for gold in gold placers; hunting for gold and silver bearing ledges or veins or deposits.

"Placers" are places where gold having been torn from the ledges and rocks by denudation, by water and ice, is swept down by these agencies till it finally finds a resting place. Gold being heavier than quartz or country rock, sinks to the bottom first. If the stream is violent, it will carry the gold on, if fine, till it comes to an eddy or pool, where the waters are more quiet, and there it will sink. The water carries the clay and lighter stones still further on. In this way millions of tons of rocks containing more or less gold disseminated through them may have been reduced, and the gold set free, or the gold may have been derived from a few individual gold bearing ledges or veins.

The prospector takes his pick, shovel and pan, and his horn spoon, and finds perhaps an old dry river bed where the water has ages ago receded. At some point the sides of this old river course widen out suddenly, forming a basin. "Here," says the prospector, "there must have been an eddy," and he prospects it accordingly; at another point he finds a place where the water must have run over a rock, and made a waterfall; at the bottom he digs again.

He loosens the soil with his pick, and shovels it out; at a

certain depth, which may be from 5 to 20 feet or more, he strikes "bed rock," which may be granite, shale, sandstone, or some other rock. Here he looks for nuggets, and with his knife digs into all the little crevices of the rock to hunt for them and for scales and wires of gold.

PANNING GOLD AT CRIPPLE CREEK, COLORADO.

Also whilst sinking his shaft, he pans the gravel carefully at various depths, especially where there are streaks of clay or "black sand." The latter are grains or little pebbles of magnetic iron ore, a common accompaniment of gold, altered relics of the iron pyrites in which the gold was originally contained.

He fills his pan half full of water, throws into it a shovelfull of dirt, first picking out the pebbles, stirs the mass with his fingers till the water is fully charged with the clay and gradually winnows out all the clay. Filling the pan again with water, he gives it a peculiar circular motion and each little wave of sand passes off till the whole is winnowed off, and at last he sees specks of gold shining free in the bottom of the pan. Then it is not difficult to estimate approximately the amount of gold to the bushel or cubic

FINDING THE FLOAT.

foot of earth of the placer, and thus to estimate the approximate value of the placer. He then locates or stakes out his placer claim according to the regulations of the U. S. Government, which, by a single individual cannot exceed twenty acres.

The second class of prospectors are those who try to discover ore deposits, ledges or veins, "in place," that is, in the hard rocks of the hills.

The prospector's first effort is to find "float." A vein outcropping on the surface, becomes oxidized and crumbles by action of the atmosphere, rain, etc.; pieces break off and fall down hill. Some of this float is barren quartz or country rock, others may be mineralized. Commonly "float" is a rusty, spongy mass of rock, showing besides iron often some copper stains, and in it there may be grains of galena, pyrite or some other ore, He tries to trace this "float" to its home in the ledge whence it came. Of one thing he is certain, the "float" must have rolled *down* and not *up* hill. If the "float" is fairly scattered over the lower zone of the hill, and no "float" is found above that zone, on the top of that zone he will hunt for his ledge. If the "float" is all over the hill he assumes the ledge is on the top.

If he finds his "float" at the mouth of a canyon or water course, he walks up that water course, noticing not only the "float," and its diminishing or increase, but also any peculiar rocky pebbles, such as a peculiar porphyry, perhaps, which he may by chance recognize again further up in place, and give him a hint as to whence the stream derived most of its material of pebbles. He notices if the "float," fragments increase as he proceeds, and whether they suddenly cease at a certain point; at that point he hunts for the ledge on either side of the canyon, and breaks off any pieces that may look likely.

Having found the ledge and traced its croppings, he tries to find out its approximate value. This he does by breaking off at intervals along it likely looking fragments of the rock, grinding them up to about the size of peas. He mixes these well, and takes a half of them, reducing this to fine powder, and again halving it, till of the whole ledge he can carry away an averaged sample of a few ounces. He may wash this in his pan to see if there is any free gold in it; other ores he will recognize at sight. These samples he will have assayed and the returns will show the approximate value. He measures the length and thickness of the vein, and examines the wall enclosing it.

He then proceeds to locate or stake it out by measuring off a parallelogram 1,500 by 600 feet. At the corners of this, he places piles of stones, and in one or more of them places a stake of wood on which he writes his name, a description of his claim and the date. At the nearest recorder's office he files a copy of this document. He must do a certain amount of improvement work on this annually, such as digging a ten foot hole or putting up a cabin or some work equivalent to the value of $100, so as to hold it. He may also claim a mill site on non-mineral land adjacent not exceeding 5 acres. Now the property is his to do as he likes with it.

THE GEOLOGICAL TRAINING OF A PROSPECTOR.

One of the first things for a prospector for gold and silver to acquaint himself with, is the elements of geology. He can read this up theoretically in many excellent treatises and manuals, such as LeConte, Dana, and Shalers' Manuals, and Geikies' Hand-Book of Field Geology, etc., and become learned in the names of eras and epochs, and the jargon of scientific names of fossils and minerals, and varieties of rocks; but let him not imagine at the end of this process, that he "*knows* geology."

Geology can no more be learned by means of a book, without field·work and the actual personal contact with nature and rocks, than chemistry or assaying can be acquired without ever using a test tube or a cupel. The student may, perhaps, be unfavorably situated for this practical field work. There may be no mountains or upheavals of strata, or deep natural ravines within available distance to study. He is located, perhaps, on the great, monotonous, flat prairie. Very well, then let him study what lies nearest him. This same flat, monotonous prairie has an interesting and wonderful history. Let him read up what he can find about this in his books, then go out and examine what he can of the few feet of horizontal strata exposed in some shallow water-course or dry ravine; examine minutely, both with eyes and microscope, the minerals composing these strata. Let him classify and collect and note the different kinds of pebbles scattered over the surface, or in the bed of a brook. Let him speculate as to the cause of the undulations of the surface, the deposition and peculiar character of the clays forming the soil. Let him study thoroughly the geology of his native village, his immediate surroundings, *first*. The knowledge and practical habit of observation so acquired, will lead later to more extensive studies in wider fields. A student may be shut up in a big city; let him study the paving stones of the streets and visit the stone yards of the masons. It will pay him better to take a trip to some distant mountain region, than to buy another expensive book on geology after he has mastered the first bare elements. Nothing like field work, eye practice, and hammer practice. The student should endeavor, whenever he possibly can, to verify by actual vision and personal experience whatever he reads in his books. When traveling, let him always carry a geological hammer with him, and at any

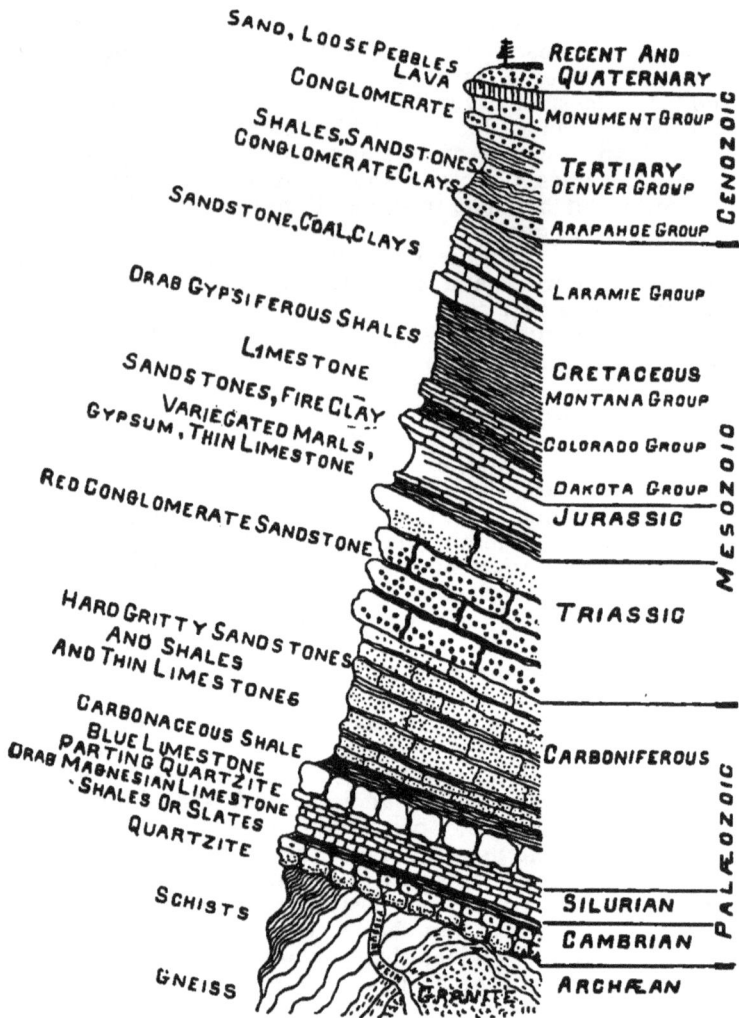

SAND, LOOSE PEBBLES
LAVA
CONGLOMERATE

RECENT AND
QUATERNARY

MONUMENT GROUP

SHALES, SANDSTONES,
CONGLOMERATE CLAYS

TERTIARY
DENVER GROUP

SANDSTONE, COAL, CLAYS

ARAPAHOE GROUP

CENOZOIC

DRAB GYPSIFEROUS SHALES

LARAMIE GROUP

LIMESTONE
SANDSTONES, FIRE CLAY
VARIEGATED MARLS,
GYPSUM, THIN LIMESTONE

CRETACEOUS
MONTANA GROUP

COLORADO GROUP

RED CONGLOMERATE SANDSTONE

DAKOTA GROUP

JURASSIC

HARD GRITTY SANDSTONES
AND SHALES
AND THIN LIMESTONES

TRIASSIC

MESOZOIC

CARBONACEOUS SHALE
BLUE LIMESTONE
PARTING QUARTZITE
DRAB MAGNESIAN LIMESTONE
SHALES OR SLATES
QUARTZITE

CARBONIFEROUS

SCHISTS

SILURIAN

CAMBRIAN

GNEISS

ARCHÆAN

PALÆOZOIC

GRANITE

PLATE I.

A Vertical Section of the Earth's Crust in Colorado.

station the train may stop for a few moments, step out and try to get a specimen of the country rock ; at the same time let him study all he can of the geology of the country he is passing through from the windows of the train, aided perhaps by a geological map. The genuine prospector is always looking about him, is everlastingly cracking stones, has always his eye wide open for "something kind o' curious."

If he is near some mountain region, where, as in Colorado, the whole strata of the earth's crust is upheaved and exposed, along the mountain flanks, in the depths of the canyons, or on the summits of the peaks, after studying his manual, let the student get, if he can, some published geological report on such a country, such as those of the U. S. Geological Survey, abounding in illustrations and geological sections. Let him take this book in hand and go to the very place described and pictured as a geological section, and with his hammer study each member of the section closely. This will make him familiar with the different geological periods, formations, rocks, minerals and fossils, *as they actually appear in nature* rather than as his *imagination has supposed* them to be from his study of the text books: book geology and field geology are not always in perfect harmony.

Having studied and learned one local section well, such as that cut by a stream along the foothills of a mountain range, let him repeat the course at the other and more distant points. He will find at each locality, though the main features are the same, there is always an interesting variety, such as new fossils, peculiar minerals, changes of dip, faults, or other structural peculiarities.

Along the flanks of a mountain range, a prospective prospector cannot study too many of these geological sections. Having become familiar with these foothill sections, he is prepared to plunge into the heart of the range itself. At first, and for long distances perhaps, he will encounter only granitic rocks forming the axis and core of the range. These are well worthy of study and full of variety. Later the canyon may open into some mountain valley or park, where the strata he studied on the foothills or prairie border are again repeated and he finds himself again at home. Seizing upon some well defined and familiar representative of a geological horizon, from this as a standpoint, he soon reads off the succession of the rest. Here, however, the. appearance and texture of the rocks will probably be different to what they were in the foothills. Heat has so changed or metamorphosed the sandstones and shales, that they are scarcely recognizable as the same rocks as those of the foothills. Yet even here a highly silicified fossil shell, or a leaf impression on shales, or sandstones changed into slates or quartzite, will give the prospector his clue and his desired and definite geological horizon, and he will have little difficulty in again arranging and grouping correctly the rocky series. But a prospector has a "practical end" in view. He is "after the precious metal," gold and silver, not after "pure science" or "fossils or sich"; what practical use in there, he may ask, in this same careful study of geological sections, where probably there is not a speck of gold or silver? Simply that minerals and metals of economic value, such as gold and silver are more frequently found in the rocks of certain geological periods than in others. Locally this is especially true. For instance, nearly all the silver-lead deposits of Colorado are found in a certain bed of limestone not over 200 feet thick, to be found only in one geological period out of many others, viz.: the lower division of the Carboniferous. It would naturally then be advisable for a Colorado prospector to be able surely to identify this limestone, as well as the geological horizon in which it occurs, among the various other limestones of various other periods and ages in the mountains.

Again, gold is mainly confined to crystalline rocks of Archæan age or to porphyries associated with these. A prospector should be familiar with these rocks and their varieties. Gold is also found in the placers derived largely from the breaking up of these rocks; the ability to distinguish the different pebbles may lead to the source whence the gold was derived. Familiarity with rocks of all kinds is a necessary prospector's education in itself.

GEOLOGICAL SECTIONS OF COLORADO.

In illustration of what we have said, let us take the two engraved generalized sections showing all we know of the crust of the earth as exposed in Colorado. Plates I and II. Plate I is a vertical section of an ideal cliff, showing all the members of the various periods in a stupendous cliff resting on fundamental Archæan granite at the bottom of a canyon. Plate II represents the same rocks and succession of strata displayed in upturned "hog backs" along the

flanks of the mountains and foothills on the border of mountain and prairie. Both of these are ideal sections "generalized" or "made up" of actual partial typical sections found in different localities in Colorado, the vertical one in detached and sometimes widely separated districts in the heart of the mountains; the other at similarly distinct and different localities along the banks of the various rivers issuing from these canyons in the mountains, cutting their way through the upturned strata of the flanking foothills and debouching on the prairie.

It is very rare to find at one locality anywhere in the world, a complete section of the earth's crust exposed. The nearest approach to this in Colorado, is the remarkable section between Colorado Springs and Manitou, which shows along the wagon road the succession of strata from Archæan to Quaternary.

One of the most remarkable *vertical* sections in the world, is in the grand canyon of the Colorado River, where the stupendous cliffs show in one face, a thickness of some 6,000 to 7,000 feet of strata, representing several geological periods, but by no means a complete section of all that is known of the earth's crust.

To show how difficult and rare it is to to get a complete section of *all* the periods in the earth's crust, we may state that sometimes the rocks of a single geological period are from 10,000 to 20,000 feet thick. A canyon might thus be cut to a depth of 5,000 feet, and yet be in only part of a single earth-period.

By far the most extensive and available sections are, like those represented in the engraving, along the courses of streams on the flanks of a mountain range. It would be a formidable task to scale a cliff 5,000 feet high and examine minutely, in ascending, each of its geological divisions; whilst, on the other hand in the foothill regions, a prospector may walk over and mark and study as much as 10,000 to 40,000 feet of strata along the banks of a river in a single afternoon. In the Weber Canyon in Utah, as much as 40,000 feet of strata, composing the flanks of the Wahsatch Range, can be seen by the traveler from the windows as he glides through in the railway car, and the inquiring prospector or geologist can examine and study this vast section leisurely on his mule or on foot, without doing any climbing and on a good road. Smaller partial sections can be similarly studied along many of the streams issuing from the Rocky Mountains among the foothills of Colorado. Such, for example, as at Boulder Creek, Clear Creek, Bear Creek, the Platte River, and, most complete of all, the one along Fountain Creek, near Colorado Springs, which we have already mentioned. Similar sections can be found in most mountain regions, such as the Adirondacks in the East, and the Sierra Nevada and Coast Range in the West of America. We emphasize again, that the close study of these is the best preliminary step we know of in a prospector's geological education. Let us now examine our ideal generalized Colorado section which we will suppose to be all exposed along the banks or canyon of a single river. We will start from the Archæan granite in the canyon, thus giving us a sure and known and lowest possible geological horizon to begin with.

THE ARCHÆAN.

This Archæan we find to be composed towards its core, of solid, shapeless (amorphous) crystalline granite, which seems to have been fused out of all shape by water and fire, or aqueo-igneous fusion. With this, but more characteristic of the upper and outer edge of the Archæan, the granite assumes a more stratified and bedded character, which we designate as "gneiss" and interbedded with it at intervals are distinctly laminated or finely leafed strata, called schist; all these varieties are composed of the same minerals in different arrangement and quantity, viz., mica, quartz, hornblende, and feldspar. As these rocks are semi-igneous or metamorphic, we find no fossils in them. Traversing all these Archæan rocks and cutting them at all sorts of angles, we may notice some eruptive dykes of porphyry, which were once certainly molten and have ascended in that state through fissures opened in the rocks from depths and sources unknown. As we approach the edge of the granite we may even see some of these molten rocks, insinuating once fiery tongues among the weak places and bedding planes of the overlying sedimentary strata, as represented in the diagram, where one dyke is shown to have sent out so thick an intrusive sheet of porphyry, (see Plate II), between the overlying limestones, that where subsequent erosion took place, this thick sheet, by its superior hardness, was left to form the highest cap of the mountain, as on many of our prominent mountain peaks such as Mt. Lincoln and others in South Park.

Besides these rocks, the prospector will observe numbers of quartz and pink feldspar veins of all sizes, some mere

streaks and occupying incipient fissures or weak places
(veins of segregation), others occupying large well defined
fissures or jointing planes (so called true fissure veins).
Some of these may or may not carry metal, gold or silver,
lead or copper, at any rate he will pay them especial attention particularly if any of them look at all decomposed or
rusty, or are in close proximity to an eruptive porphyry
dyke.

THE CAMBRIAN.

Now the prospector emerges from the Archæan granite
and finds the first true sedimentary, water-formed rocks
lying where the ancient seas placed them, on the eroded
upturned edges of the granitic series.

If this section should be near the plains or foothills,
this first sedimentary rock will be a sandstone, pure and
simple, or a conglomerate of little pebbles, but in the parks
and center of the mountains where these ancient strata are
most conspicuous, the first rock lying on the granite is a
hard, white, semi-crystalline quartzite or metamorphosed
sandstone. He may possibly find some obscure signs of
ancient fossil shells in this series, which is called the Cambrian now, though formerly it was held to be only a lower
division of the Silurian. In Colorado these Cambrian rocks
rarely exceed 200 or 300 feet in thickness, but in other
regions they are often very much thicker. In this series
the prospector may look for precious ore, more especially
gold. He will carefully look also for intrusions of eruptive
porphyry in this series, as at the junction of this with the
quartzite, ore is most likely to be found. He will also
observe any rusty signs filling cracks, as good indications
of gold bearing ore. Silver also may be found associated
with lead or zinc.

SILURIAN.

Walking along, he next comes to some 200 or 300 feet of
drab-yellowish or light gray thin bedded limestone of a
dolomitic character, characterized by numbers of little white
flints or (rarely in Colorado) by some fossil shells, which, by
reference to the engravings in his manual, he finds to be
Silurian, and so recognizes the series. Here he may find
indications of lead, silver or other ores, but not much gold
as a rule.

CARBONIFEROUS.

The next series of this should, according to the text-
books, be the Devonian, characterized by fossil fishes and

"Old Red" sandstones; but the rocks of this epoch for some reason are missing in Colorado. Instead of this, resting on the Silurian, he finds a thick bed of heavy bedded, massive, "blue-grey" limestone, characterized by black flints, and at rare intervals by fossil shells and corals, which again, by reference to his book, he finds to be characteristic of the Lower Carboniferous. This limestone when traversed by sheets of eruptive porphyry, has yielded at Leadville and at Aspen and New Mexico and Arizona, some of the largest silver-lead deposits in the West. In fact, throughout the West it may be considered as the main silver-lead horizon. This limestone is generally between 200 and 300 feet in thickness and readily recognized by its position relative to the Silurian below it, and the massiveness of the strata, and their dark grey color. It is commonly called the "Blue Limestone" in Colorado.

MIDDLE CARBONIFEROUS.

Next on this, is a bed of dark black shales in which thin seams are sometimes found, and fossil plants, like those in the coal strata of Pennsylvania, sufficient to show that it, too, belongs to the Carboniferous. This is followed by some 2,000 or more feet of "grits," rough, hard, gritty sandstones, partially changing into quartzite, akin to the "mill-stone grits" of the Eastern States. A few limestones occur in this thick Middle Carboniferous series, which locally, when capped by porphyry, produce silver-lead deposits; but generally speaking, the "grits" are unproductive in Colorado.

The Upper Carboniferous consists of beds of gypsiferous shale and heavy, brownish red conglomerate sandstones.

TRIASSIC "RED-BEDS."

From these we pass into a series of heavy bedded, coarse conglomerate sandstones of a brick-red color, commonly known as the "Red Beds" in Colorado; little indications of ore are to be expected in this series. The prevailing redness of the series makes it an easily recognized geological horizon in Colorado and elsewhere. The thickness in Colorado varies from 1,000 to 2,000 feet.

JURASSIC.

Next, the prospector comes to a softer and more variegated series, consisting largely of pink, green, red, or maroon marls and clays, with some thin limestones and red sandstones. This is the Jurassic series in which some remarkable lizard remains, called Dinosaurs, have been found, proving the correctness of its Jurassic name. This is not a likely mineral horizon, generally speaking, in Colorado.

CRETACEOUS.

These softer beds are capped by a hard massive sandstone about 200 feet thick, forming by reason of its superior hardness a prominent hog back in the prairie or foothill region. Fossil remains of leaves show it to be a land and fresh-water group, which is called the Dakotah group.

This group in Colorado forms the base of the great Cretaceous system; lying on it, is an enormous thickness of drab shales with a few limestones characterized by fossil sea shells, showing the group to be the marine Cretaceous, likewise a poor prospecting ground. Towards the upper portion, these shales pass gradually into heavy bedded sandstones containing several seams of coal, and many impressions of tropical foliage. This is the Laramie group of the Cretaceous, evidently of fresh water origin, and noted as the main coal producing horizon in Colorado and the West.

TERTIARY.

On this, at a somewhat gentler angle even to horizontality, rest thick beds of shale and clay and conglomerate, composed of volcanic detritus and pebbles, showing that at the time these Tertiary beds were being laid down by large fresh water lakes and marshes surrounded by tropical foliage, volcanic eruptions on a grand scale repeatedly occurred. Hence it is that many of the Tertiary beds are preserved from erosion by being capped with volcanic rocks, such as basalt, andesite, or rhyolite, as at the Table Mountains at Golden, on the Divide near Colorado Springs, and elsewhere in Colorado. One of these lava capped "mesas" is represented in the section, Plate II. Fossil leaves and coal seams are found in this period.

QUATERNARY.

Lastly, strewn indiscriminately over all the formations is the "Quaternary drift" composed of loose pebbles, and sands, and clays, the material derived from rocks of all the periods through the agency of glaciers and streams.

Here the prospector will pan for his gold placer, and in his search may possibly come across the teeth or tusks of the great Mammoth or fossil elephant, together with the first indications of the presence of primitive man. The pebbles by their variety will form a fertile subject of study to determine to what class of rocks they belong.

This ends the prospector's first preliminary lesson in Colorado; but taking this section as a type, he may to his great advantage, similarly study other sections far remote from Colorado.

In Colorado, if he knows this section by heart, he has the key to nearly all our mountain structure, and will be at home wherever he goes. He will be struck, too, to see to how small a portion of this great section the precious metals are more or less confined, principally to the Archæan and Paleozoic rocks.

CHAPTER II.

THE PROSPECTOR'S HISTORICAL GEOLOGY.

In our last chapter we gave some hints to the prospector how to commence his geological studies, and gave him an example of a geological section of the foothills and mountains of Colorado, and how to study it in detail practically. Having completed this study, if a thoughtful man, he will like to know more of the natural history of all this section of the earth's crust: what is the natural history of the Archæan, the Cambrian, Silurian, etc., why do some of these strata contain sea shells, and others land plants, why are some evidently of marine, and others of fresh water origin, and particularly why are some especially metalliferous, and others not so much so. We propose, therefore, in this chapter to give him a brief sketch of the earth's history as exemplified in the section, Plates I. and II.

HYPOTHETICAL ORIGIN OF THE EARTH.

The world was not "spoken into existence ready made" in the state we now find it. It has attained this condition through a multitude of gradual changes and revolutions which have taken millions of years to accomplish. The remote history of the earth's origin is a matter of hypothesis and speculation. There are reasons for supposing that at one time its elements were in a gaseous condition, and that this planet was an incandescent luminous cloud revolving through space, gradually consolidating into a molten ball surrounded still by an atmosphere of gases, a condition perhaps not very unlike that of the sun, whose interior by some is supposed to be passing into the molten state, while its exterior consists of various incandescent gases arranged more or less according to their specific gravities. The spectroscope has detected the elements of some of our earth metals and minerals in the sun in a state of vapor. The ultimate source of the precious metals is again a matter of speculation like the nebular hypothesis we have alluded to, by which the earth, as we have said, is supposed to have arrived at its present condition as the result from the gradual cooling of an incandescent mass, and as the specific gravity of the crust is much less than that of the whole mass of the earth, it has been inferred that the heavy metals must be in much larger proportion in the interior of the earth, than in the rocky crust, though this greater interior specific gravity might be also accounted for by the rocks of the interior being much more tightly packed by enormous pressure than those near the surface. Volcanic emanations and hot springs contain metallic minerals, so also do the waters of the ocean. But we know not from what depth the former came, nor from what source the latter derived them. As circulating waters take up and throw down their metallic contents under varying conditions, the same material may have been deposited more than once, and in more than one form since it reached the rocky crust.

Upon the cooling of the ball, a crust formed like that on molten iron, crumpled and corrugated by contraction, due to cooling, into an uneven surface, with comparatively slight elevations and depressions, and doubtless broken through here and there by great fissures and volcanic craters, through which the molten flood beneath poured out in volumes, adding to the thickness of the congealing crust.

Upon such a surface the gaseous atmosphere, gradually cooling and condensing, descended as hot chemical rain, and filled the troughs of the crumpled surface with a hot, chemical, steamy ocean. Whatever land of primitive lava rose above this ocean was battered by the waves, reduced to sediment, and deposited as the first sedimentary strata in the bed of that primæval ocean, the eruptions from below the thin crust doubtless contributing largely to the same material.

ARCHÆAN AGE.

Thus, perhaps, were formed the first stratified rocks of the world, which we have an opportunity of actually seeing and studying, viz.: the granitic series, with its varieties of gneiss, schist, syenite, etc., and as this is the beginning age so far as we know, we call it the Archæan, the Greek for beginning. It would seem probable, however, that these granitic rocks forming the axes of our mountains, may not, at least in part, have been the very first rocks of the crust, for we observe some of them such as the gneisses and schists to be stratified, and to show elements in them seemingly derived from other and still older rocks, which latter may or may not have belonged to the original cooling crust. Some geologists claim that the Archæan is the first cooled crust and attribute it to a molten origin. This may be true for the seemingly fused massive amorphous granites (though these may be but the result of aqueo-igneous fusion of sediment or extreme metamorphic action), but scarcely for the stratified gneisses and schists, though it is to be noted that a sort of stratified or schistose structure is sometimes observed in truly igneous rocks and may be induced by peculiar arrangement of minerals, pressure and cleavage, instead of water lamination.

The subject is a difficult one and too abstruse for the limits of this work.

In the scale of geological periods in the text-books, we sometimes find this great Archæan divided into two or more groups such as the Laurentian, Huronian and of late the Algonkian. The Laurentian is the oldest and may be called the Archæan proper, whilst Huronian and Algonkian may be grouped generally as Pre-Cambrian, or series of rocks laid down after the Laurentian and before the Cambrian. All the rocks are of a highly crystalline order and have a peculiar and distinct general appearance different, as a rule, to those of any subsequent geological periods and so not easily mistaken for them, consisting in the lower division, mainly of granite, gneiss and schists, and in the upper divisions of gneisses, schists, quartzites, slates, some marble, serpentine, etc. The upper or Pre-Cambrian series is not nearly so universally found as the Laurentian or Archæan proper. In Colorado we find the Pre-Cambrian represented locally in South Boulder and Coal Creek canyons, along the foothills, also near Salida in the Arkansas valley, in the Quartzite range and on the road between Ironton and Ouray in the San Juan Mountains. The new Kootanie silver mining district of British Columbia, seems to be largely in these Pre-Cambrian rocks. This Pre-Cambrian is usually very thick, numbering many thousands of feet.

It is distinct from the Archæan proper or Laurentian by lying on the latter at a different angle, in other words "unconformable." The rocks, too, do not contain so much of the heavy massive granites, and heavy bedded gneisses as the Laurentian, but are more characterized by quartzites, by conglomeratic gneisses and schists, and show clearly that though highly metamorphosed and crystalline, they

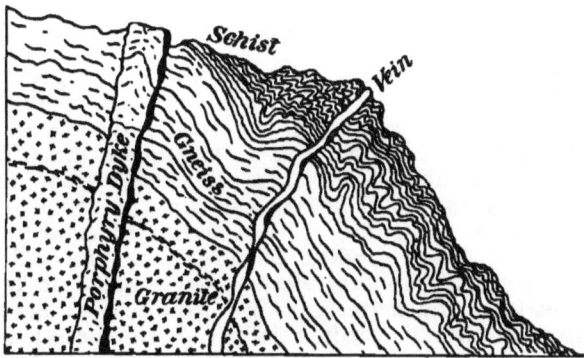

PLATE III.

Archæan Rocks.

are of true fragmental and aqueous origin, for the pebbles in the gneiss are often very distinct, and ripple marks are not uncommon on the quartzites and slates and schists. The material was doubtless derived by waters from that of the underlying and older Laurentian. The whole Archæan series, however, has evidently passed through an ordeal of heat, such as is called aqueo-igneous heat, and all its elements are in a highly crystalline condition. Its strata are intensely folded and crumpled. See Plate III.

Signs of life, in the upper series even, are exceedingly obscure and doubtful, such as graphite and possibly corals. Great iron beds also occur, indirect proofs perhaps of the previous existence of life.

We have been thus particular with this Archæan Age because its rocks are of great importance to the prospector, being the main repositories of gold, silver and the precious metals thoughout the world. Moreover many of the other and newer rocks containing gold and silver have been made from the detritus of this, and the gold placer beds largely from the detritus of the rocks and porphyries found in this age. Thus the Archæan may be considered as the parent of nearly all the other rocks. When later we have studied the origin of ore deposits, we shall see how eminently the Archæan Age with its attendant heat, chemical reactions, fissuring, metamorphism and volcanic eruptions was favorable to the diffusion and concentration of precious ores in its rocks.

CAMBRIAN AND SILURIAN AGES.

Cooling and consequent contractions still progressing in the globe, fresh and greater wrinkles and corrugations were caused on the surface of its crust, and some of these granite seabottom strata were crumpled up, till the crumples arose above the then universal ocean as low islands or reefs. The ocean had by this time cooled sufficiently to support low forms of marine life, and so along the flanks of these granitic islands, corals formed reefs, shell fish swarmed and sea weeds grew. Sands formed by the waves from the material of the granite were laid down as shore-line beaches, often mixed with shells; and in deeper water, corals were forming limestones as at the present day, both, by time and pressure, consolidating into hard rock, eventually it may be, metamorphosed by heat into a semi-crystalline hardness, as in the case of the Cambrian quartzite and Silurian limestones, the latter sometimes changed to marble. If these Cambrian quartzites were formed from the detritus of the granite and the granitic series is the source of gold, it is not surprising that we find the Cambrian quartzites *locally* rich in gold, as they were the auriferous sea beaches (like those of to-day in California which are gold bearing) of that period, later consolidated into hard rock. In Colorado the Cambrian quartzites are only locally prolific in gold, as at Red Cliff, but as they have hitherto been much overlooked by prospectors they are worthy of closer attention by the gold seekers. The limestone not being. of a true fragmental origin but formed by the slow work of corals, could not be expected on consolidation to be a recipient of gold, but later by its peculiar chemical composition, of which we will speak hereafter, and by its cavernous nature, it furnished a more convenient receptacle for silver and lead ores.

So then in Colorado and in other regions, we find first the upheaved crumpled granite of the old Archæan island, and on these the Cambrian sandstone or quartzite beach of "golden sands" with some fossil shells, and upon this again Silurian limestone with relics of fossil corals and shells. So we call these ages the Cambrian and Silurian because the fossil shells and corals are peculiar to those ages and distinct from those of later periods or the present day.

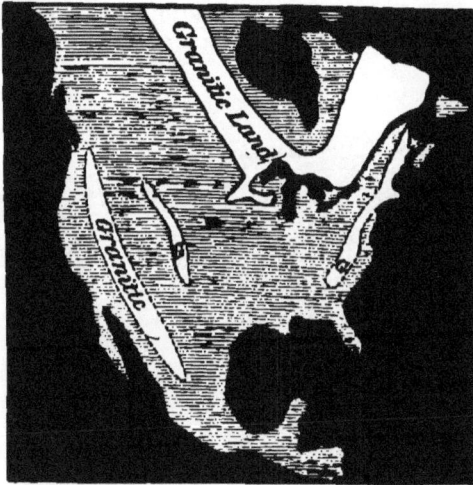

PLATE IV.

America at Close of Archæan.

North America at the beginning of these periods was barely outlined by a few granite islands congregating mainly in the region now occupied by Canada, whilst one or two reefs or scattered chains of islands marked the site of the Eastern ranges of mountains, and a few parallel granite islands outlined the site of the principal uplifts or future great ranges of the Western Cordilleras. All else was ocean, and that ocean was depositing its Cambrian beaches and Silurian coral limestones against or near these granite islands destined in time to grow into lofty mountain ranges, and to become the backbone of the American Continent. See Plate IV.

DEVONIAN.

The Devonian which should come next in order in the geological tree appears to be absent in Colorado but is well shown at the Eureka Mines in Nevada. The rocks appear to be mostly marine limestone full of corals and shells and a few remains of gigantic fishes for which this age was celebrated. Land plants and some coal are found in it in the East. Lead silver ores may be expected in the limestones of this age, and in Cornwall (England) Devonian slates traversed by quartz porphyries are the main rocks carrying tin ore, a metal very scarce at present in North America.

PLATE V.

Section showing Unconformity of Geological Eras.

These ages we are speaking of are separated or distinguishable from one another by decided and characteristic changes in the fossil, animal and vegetable life existing between one age and another, also in some countries by marked unconformability of the rocks, *i. e.*, the rocks of one age lying at a different angle upon the upturned rocks of a previous age marking great oscillations between sea and land.

In America, however, these oscillations between sea and land seem to have been less than in Europe, and we find a general uniform rise of the continent from the primitive oceans, and an orderly succession of strata lying against the flanks of the ever rising granite nucleus of both mountains and continent. Hence to distinguish the different ages we are driven more to the study of fossils and lithological peculiarities than deriving any help from observed marked unconformability. See Plate V, in which the strata of the different eras lie upon one another at different angles, and the glacial and Quaternary drift pebbles and clays are strewn unconformably also over the tops of the uptilted and eroded strata of all the eras beneath.

CARBONIFEROUS.

In the Eastern States as the American continent gradually rose from the sea, and to the granite islands had been added a Cambrian, Silurian, and Devonian shore, with further unequal elevation, a kind of wide trough or synclinal fold or depression appears to have been formed between the middle and eastern part of America, which was at first occupied by a wide arm of the sea, later, by continued elevation, by a great body of fresh water, and later by low marshes and low marshy islands barely above sea-level. Upon these low lying lands grew a dense vegetation unlike any of the present day, but resembling somewhat the tree ferns of our southern semi-tropical States. This low lying region was subject to freshets and inundations from the surrounding higher regions, periodically deluging the swamps and swamp vegetation with river and flood deposits of pebbles and sand, under pressure of which the peat gradually turned into coal. Successive coal seams were formed by successive growths of vegetation between the intervals of periodic inundation, or of subsidence and possibly at times of upheavals, for these low lands, as sediments accumulated, appear at times to have sunk below the sea and again to have been either built up above it by fresh supplies of sediment, or to have been temporarily raised up by upheaving forces.

Finally by a grand revolution which closed the Carbonif-erous age in America, the coal swamps with their coalbeds and strata were crumpled up to form the present great Appalachian Chain.

Similar movements no doubt took place about the same time in the Rocky Mountain and Western region. But here the marine condition seems to have predominated over the fresh water one, for we find the Carboniferous in Colorado more represented by marine fossiliferous lime-stones and sandstones than by those of fresh water origin, though the Weber-grits may have had a fresh water origin, as in a few rare instances we find fossil plants like those in Pennsylvania together with a few insignificant small seams of coal. But in the West it is evident that the circumstances from one cause or another were not favorable for the production and growth of extensive coal-beds as in the Eastern States. The coal forming time was reserved in the West for a much later period, viz.: the Laramie or Upper Cretaceous. The Lower Carboniferous in Colorado, however, contains in its limestones much of our silver-lead wealth as at Leadville and Aspen.

The Cambrian, Silurian, Devonian, and Carboniferous Ages have been grouped together by geologists into one great era, the Paleozoic, owing to a general family likeness in the fossil fauna and flora of these ages.

To the Archæan and Paleozoic rocks the bulk of our veins and deposits of gold and silver are mainly confined, though both in Colorado and elsewhere, as will appear later, if certain peculiar conditions are present, the rocks of the later and newer periods may also in some regions produce precious ores. But the prospector should give his closest attention to these *older rocks*, hence we have devoted extra space to their description and history.

TRIASSIC AND JURASSIC, OR JURA-TRIAS.

After the Carboniferous, followed the Triassic and Jurassic; sometimes in America, owing to the difficulty of positively separating the two periods, they are combined under one name, the Jura-Trias, and in Colorado are locally called the "Red-Beds," owing to their prevailing red and variegated colors. The series is well represented in the celebrated Garden of the Gods, near Colorado Springs. The red conglomerate sandstone of the Trias proper, has so far yielded no determinative fossils, but the variegated clays in the upper Jurassic at Morrison and elsewhere have yielded some remarkable Saurian remains of land lizards. It is probable from the presence of salt and gypsum in these red-beds, and the prevailing redness of the rocks, due to iron, which was not leached out through the agency of organic life, and the general absence of fossil remains, that the lower portion of these rocks was laid down in land-locked salt seas, or salt lakes, shunned by both vegetable and animal life. The upper portions, however, show evidence of the existence of land of a low marshy character, with fresh water and probably large estuaries, as we find the remains of turtles, crocodiles, fresh water shells and Dinosaurs or land lizards. The rocks of these periods are not generally prolific in ores. The Silver Reef sandstone of

33

Utah is an exception, which contains chloride of silver disseminated through it. When pierced by eruptive rocks, however, ore should be looked for in this series as elsewhere.

CRETACEOUS PERIOD.

Upon this followed the Cretaceous, a series of very thick formations, numbering several thousands of feet in Colorado, consisting in its middle portion of limestones, and thick beds of drab shale. These are mostly marine, as shown by the sea shells in them, but at the base is what is called the Dakotah group or Cretaceous No. 1, a prominent

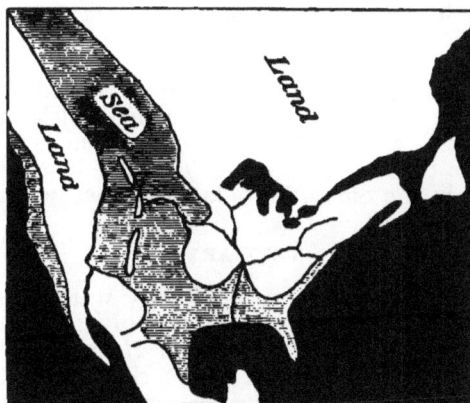

PLATE VI.

North America in the Cretaceous.

sandstone hogback in which the fossil impressions of leaves, very like, but not identical with those of the present day, show that land and fresh water existed at the time. The limestones and clays of the middle or Colorado group, contain quantities of fossil marine shells, such as the Nautilus, Ammonite, Baculite and Inoceramus.

The Laramie forms the upper group of the Cretaceous, and contains our principal western coal fields and abounds in fossil remains of tropical foliage.

This Laramie group marks an important era in our Rocky Mountain region for it shows that beginning of the great Rocky Mountain revolution, by which the granite islands before mentioned, against which all the previous sediments had been forming mainly beneath the sea, were elevated 10,000 feet or more into continental or mountainous masses, dragging up with them portions of the sea bottom and exposing it as land surface, draining off the shallow Cretaceous sea which had hitherto divided the Eastern half of the American continent from the Western, bringing on a land and continental condition, which was completed in the following Tertiary age and has continued to the present. See plate VI.

The Jurassic, Triassic and Cretaceous are grouped into one main division called the Mesozoic or middle life era of the world's history. None of the rocks of this age in Colorado are celebrated for ore deposits, except locally under local conditions.

In California and portions of the extreme West where these rocks have been highly metamorphosed by heat and penetrated by igneous rocks, some of the leading ore deposits of gold and silver are found. The same remark applies also to the succeeding Tertiary in those regions, particularly in the Sierra Nevada and Coast ranges.

TERTIARY.

The Tertiary age seems in the Rocky Mountains to mark an era of comparative rest in mountain elevation, for the strata forming some of the divisions of this age lie almost horizontally upon the tops of the earlier upturned periods.

These beds were formed by fresh water lakes in Colorado surrounded by tropical vegetation. In the Coast ranges of California the Tertiary is upturned into mountain forms and metamorphosed, and, from the presence of sea shells, is clearly of marine origin. The Tertiary in Colorado is best seen in outlying table lands. In Wyoming the Tertiary lake formed the Green River beds and Bad Lands abounding in fossil mammals, leaves, fishes and insects. The Tertiary was the world's tropical summer, a period of beautiful lakes of semi tropical foliage and a warm climate. In certain regions it was disturbed by gigantic revolutions which upheaved the Himalayas and the Alps. Such revolutions as occurred in our Western Cordillera system were marked by enormous ebullitions of lavas of various kinds issuing from fissures deluging Idaho, Nevada, part of Oregon, and Washington. Remnants of this same disturbance are seen in the form of basaltic overflows capping Tertiary strata in Colorado and New Mexico; and the vast volcanic region of San Juan in Southern Colorado is covered with successive lava overflows of the same period.

The Tertiary rocks in Colorado are not generally good prospecting grounds. The lavas, however, are (with the exception of the basalt, which for some reason is generally sterile) locally productive, as for instance the entire San Juan Region, also Cripple Creek Mining Camps and Silver Cliff. So, the prospector, whilst he need not waste time among the sedimentary beds, will do well to examine any eruptive rocks of this period for gold especially, and also for silver. The varieties of lava are principally andesite, rhyolite, trachyte and basalt. In the Coast range of California where the Tertiary beds have been metamorphosed by heat into slates, gold and cinnabar are found.

GLACIAL EPOCH AND QUATERNARY AGE.

The Tertiary Summer was closed by the world's Great Winter. The ice from the north pole for some reason we will not discuss, extended its domain far south to latitude 40. All the northern temperate regions of the world were ice-sheeted and the sheet extended itself as by long fingers down the, by that time, highly developed mountains, filling the ravines with glaciers. By the downward destructive grinding motion of the glaciers, the ravines, commenced by water, were deepened and widened by ice. Fissure veins were thus exposed, both of gold and silver. The debris from their progress the glaciers carried on their backs and dumped at the outlet of the canyons; and when the temperature finally became warmer, and the glaciers melted, all the long lines of traveling boulders scattered upon their backs, many of them containing gold robbed from the veins, were left as banks or "moraines" forming our "gold placer" grounds along the sides of our streams and canyons, or sometimes a thousand feet above the present river bed, marking the original height or thickness the great ice bodies once attained.

So were our canyons largely formed, and so did our gold placers originate. After the Glacial Epoch, a warmer period set in, called the Quaternary. The ice melted. Vast bodies of fresh water were distributed in wide streams and monstrous lakes over large portions of this hemisphere. The rough "morainal" dumps of the glaciers were "sorted." or "modified" by water, rolled into pebbles and sand, and redistributed along the banks of streams or carried out into beds of lakes. In these pebbles and sand, was much of the precious metal mined and robbed from the veins. The gold by its insolubility remains to this day in our placer beds and "drift" or "wash" and is collected by hydraulic mining. That the prospector for gold should closely study these Glacial and Quaternary deposits is evident.

So ends the history of our section. Still the agencies of nature are at work as of old. Continents are gradually rising or sinking. Mountains are being imperceptibly elevated. Water is still sculpturing them with canyons. Rivers are carrying down fragments robbed from the land and depositing them in the ocean to form strata for future continents.

The fires of the earth are not yet dead, for volcanoes still vomit lava. The earth, however, is still continuing to lose internal heat. Its crust is still contracting and wrinkling itself upwards, for we find modern sea beaches raised high on our seaboard cliffs. Shocks of earthquakes from time to time, prove that motion of some kind is going on beneath us, and doubtless our mountains are still rising imperceptibly, as they appear to have done in ages past, giving additional lifts and elevation to old uplifted strata, and slowly elevating newer strata that since the Tertiary have lain apparently undisturbed. We say apparently, for not only are the Tertiary beds uplifted from 5 to 10 degrees, but even the more recent Quaternary deposits, showing that movement has been going on comparatively recently and may still be progressing imperceptibly.

CHAPTER III.

THE PROSPECTOR'S PALEONTOLOGY OR STUDY OF FOSSILS.

A prospector in his roaming among the rocks is. likely from time to time to come across a good many fossils or petrified remains of life that once existed on this planet. He will feel curious to know what these are, what class of animal or vegetable they may represent, to what geological era, epoch, or subdivision they may belong.

CHARACTERISTIC			ROCKS.	MINERALS, METALS &c.	FOSSILS.
Orchwor		Recent	Soil, Clay, Pebbles	Some Gold	Man, Buffalo &c.
		Quaternary	Pebbles Sand, Clay	Placer Gold	Elephants teeth, Bones Man Bones, Tools
Cenozoic	Tertiary	Pliocene	Loosely Stratified Conglomerates, Sands	Old Gold Placers in California	Fossil Leaves
		Miocene	Basalt Andesite Rhyolite Lavas Conglomerates Sandstones	Lava-Gold & Silverbearing Thin Lignite Coal Metamorphosed Sandstone Gold bearing in California also	Mammals
		Eocene	Shales & Clays Some of Volcanic detritus, others of Granitic detritus	Ashphalt in Colorado and California	in California Marine Shells
Mesozoic	Cretaceous	Laramie	Clay Coal Beds Sandstones	Coal	Leaves, Trees &c. Sea Shells
		Colo.	Drab, Shales, Clays Limestone Dark Shales	Canon City Oil Horizon Flux Lime Clay, Iron, stone	Scaphites Baculites Sea Shells Inoceramus Oysters
		Dakota	Conglomerate Sandstone	Fire Clay	Leaves of trees
	Jurassic		Varigated Clays Red maris, Sandstone Lime Stone	Gypsum Oil and Lime & Red Building Stone	Dinosaurs Colo. Sea Shells in Wyoming
	Triassic		Thin Lime stones Thick Red Conglomerate Sand Stone	Copper Silica for Glass Silver, Reef Sandstone Some Red Building Stone	Foot prints of Saurians
Paleozoic	Carboniferous		Gypsiferous Shales Reddish Conglomerate Eastern Coal Beds Shales, Sandstones Grits & Shales	Eastern Coal of Pennsylvania	Land Plants
		L. Carb.	Blue Limestone	Silver. Lead	Corals Sea Shells, Spirifers, etc.
	Devonian		Reddish Sandstone Lime stones	Eureka, Nevada Silver. Lead Deposits	Sea Shells Fish Corals
	Silurian		Drab Pale Limestone Dolomite	Marble Silver, Lead Iron	Sea Shells Crustacea Trilobites Corals
	Cambrian		Slates Quartzites	Gold	Sea Shells
Archæan	Archæan	Pre Camb.	Quartzites, Conglomeritic Gneiss, Schist Slates, Marble Granite, Gneiss Schist, Syenite	Gold, Silver Lead, Zinc, Copper &c. Iron	Few Positive Signs of Life

PLATE VII.

Prospectors' Geological Table of Western Formations, Showing Principal Characteristic Rocks, Minerals and Fossils to be Found in Them.

Fossils to a geologist are the labels of the rocks; show a geologist a fossil, and he will probably be able to tell at a glance whether the fossil came from a series of Paleozoic, Mesozoic or Cenozoic rocks, whether it belonged to a very ancient geological period down near the primitive granite, or to a comparatively recent one near the modern soil, high up in the geological scale and nearer to the life of the present day. He may be able to tell not merely whether it belongs to one of the great divisions, to the great eras, but also to the subdivisions of these eras, whether to the Silurian or Carboniferous, the Jurassic or the Cretaceous, or even to minor divisions of these, called groups; whether, for example, it belongs to the Dakotah group of the Cretaceous, or to the Laramie group of the same period.

PRACTICAL USE OF FOSSILS.

The practical use of a general knowledge of fossils is obvious. A prospector finds in certain strata a fern-leaf of the Carboniferous, this tells him he must be on the coal strata and forthwith he hunts for coal. Or he finds a Paleozoic shell or coral which points to the fact that he is probably in the neighborhood of the precious ore-bearing rocks.

Later perhaps he finds a shell or coral characteristic of the lower Carboniferous blue limestone, the celebrated lead-silver bearing formation of Colorado and the West, and he is encouraged to look for these ores. The limestone by itself is but a poor guide, for there are many limestones not unlike it in the different series of rocks, but this particular shell labels this as "*the* blue limestone" and no other. Hence a characteristic fossil may help considerably in following up in its extension an ore-bearing rock, and not only that locally, but in regions very far apart. Soon after the celebrated ore deposits of Aspen were discovered, and the mines were in their infancy, some fossils were discovered that showed the deposits to be in the same limestone as that at Leadville, which had proved there so productive. This gave an additional impetus to the camp, "a second Leadville" so it was said.

Again, though a prospector may not find at once the particular geological stratum or period he is looking for, if he finds a characteristic fossil anywhere, in some other period, he knows from it whether the period he is after lies geologically below or above where he is looking.

SILURIAN.

In the next series, the Silurian, he may be more fortunate. He may find remains of sea-weeds, corals and shells and fragments of a sort of sea-worm called a Crinoid, or sea lily. The little discs with a hole in the center forming a little ring about the size of a pea, constituting the discs or rings,

PLATE IX. –SILURIAN FOSSILS.

1. 2, Orthis; 3, 4, Spirifer; 5, Pleurotomaria; 6, Murchisonia; 7a, 7b, Trilobite (Calymene); 8, Coral Fenestella; 9, Coral Choetites; 10, Graptolite; 11, Orthoceratite.

of which the stems of the sea lily are composed, are sometimes very common in Silurian and Paleozoic rocks, though it is rare to find a complete Crinoid, and especially the beautiful comb-like flower or head of the sea lily. He is likely to find also a more advanced type of the Trilobite and various Spirifers and other shells as pictured. Plate IX.

DEVONIAN.

In the Devonian he may find the teeth or bones of fishes, and a few remains of peculiar land plants, neither of which are known in the Silurian below, also many corals.

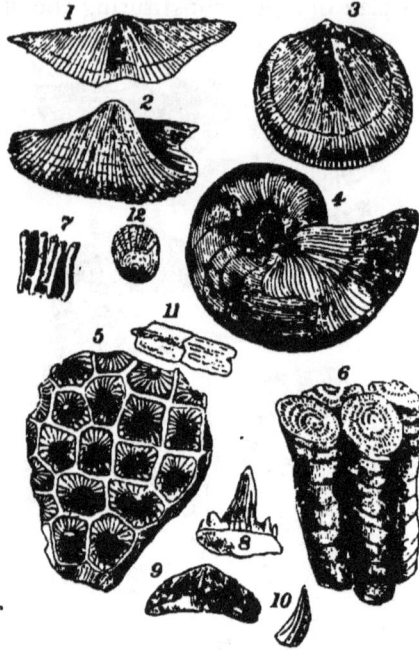

PLATE X.—DEVONIAN FOSSILS.

1, Spirifer ; 2, Comocardium ; 3, Orthis : 4, Goniatites ; 5, 6, 7, Corals ; 8, 9, 10, Fish Teeth ; 11, 12, Fish Scales.

CARBONIFEROUS.

In the Lower Carboniferous "blue limestone," corals and shells appear, especially Spirifers and Productus, together with Crinoids and a very simple curled shell like a snake coiled up, a "Goniatite," one of the earliest of the Ammonite class. At Aspen, associated with the ore deposits we found in the blue limestone most of these, together with a kind of snail shell called Pleurotomaria. At Leadville in the same

43

formation Spirifers and Productus are occasionally found. A very curious coral is one shaped like a screw, called Archimedes, after the author of the screw. Cup corals are common.

In the Middle Carboniferous, associated with the coal seams, many curious remains of reeds, ferns and other aquatic plants of that age are found, but these are scarce in Colorado and the West. The prospector will observe that

PLATE XI.—CARBONIFEROUS FOSSILS.

1a, 1b, 1c, Productus; 2, 2, Spirifers; 3, 3, 3, Rhynconella; 4, Euomphalus; 5, 5, Crinoids; 6, Pleurotomaria; 7. Bellerophon; 8, Athyris Subtilita ; 9, Astartella ; 10, Goniatites ; 11, 12, Corals ; 13, 14, 15, 16, Plants; 17, Spine of Echinus.

there is a general family likeness between the fossils of each division of the Paleozoic and in the Paleozoic as a whole, and it may not always be easy for him to determine whether a shell is Silurian, Devonian, or Carboniferous, but of one thing he will be certain, that it is Paleozoic.

TRIASSIC.

In the Trias throughout the West, he is not likely to find many fossils, the rocks are generally too coarse, but in the Eastern States, though he may not find any true remains, he may observe the tracks left by great Saurians, as they walked on their hind feet, or on all fours, on the red sands of the beaches of those dreary salt Triassic seas, leaving "footprints on the sands of time" full of interest.

JURASSIC.

In the Jurassic shales and limestones in Colorado, he may be equally unsuccessful, though in the upper Jurassic just

PLATE XII.—JURA-TRIAS FOSSILS.

1, Dinosaur Lizard; 2, 3, Foot and Shoulder Bone; 4, 4, Vertebra of Sea Saurian, Ichthyosaurus; 5, 6, 6, Teeth of Saurians; 7, Belemnite; 8, Echinus 9, 9, Ammonites; 10, Exogyra; 11, Trigonia Shell.

below the Dakotah sandstone, he may light on the bones of gigantic Dinosaurs, or great land lizards, such as the author found in Colorado and Wyoming, monsters 60 to 80 feet in length and proportionally tall, standing from 20 to 25 feet in height. In the lower Jurassic in Wyoming, he will find great numbers of sea-shells and Ammonites, and a 'round shell like a cigar called a " Belemnite " or spear-head, the internal shell of an ancient cuttle fish. Plate XII.

CRETACEOUS.

In the Cretaceous, beginning with the lowest group, the Dakotah group, net-veined leaves of deciduous trees, such as the willow, oak, maple, etc., the earliest known leaves of those kinds of trees, may be expected in the sandstone and clays.

PLATE XIII.—CRETACEOUS FOSSILS.

1, 1, Inoceramus ; 2, Cardium ; 3, Corbula ; 4, Mactra ; 5, Margarita ; 6, Fasciolaria ;
7, Anchura ; 8, Pyrifusus ; 9, 10, Scaphites ; 11, Crioceras ;
12, Baculites ; 13, Shark's Tooth.

In the Colorado group of the Cretaceous, above the Dakotah, abundance of oyster shells and large clam shells (Inoceramus) are sure to be found in the limestones and marine shales. In the Montana group of the Cretaceous above this, consisting mainly of drab shales and some sandstones, great quantities of sea-shells are found, amongst them various peculiar forms of the Ammonite allied to the modern nautilus, and called Scaphites, resembling snakes or worms uncoiling, together with shark's teeth and bones of sea Saurians.

In the sandstones of the Laramie Cretaceous remains of sea-weeds are found; in the sandstones immediately below the coal-beds, and in those associated with or above the coal, are found great varieties of semi-tropical leaves, such as those of the palmetto, fig, beech, elm, magnolia, sassafras, etc. The presence of these leaves is a pretty sure indication of coal.

TERTIARY.

In the Tertiary fresh-water beds, similar leaves and thin beds of poor lignite coal are found, together with fossil

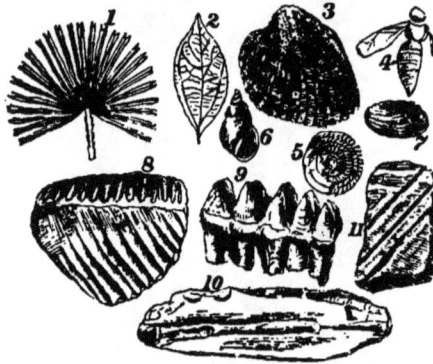

PLATE XIV.—TERTIARY FOSSILS.

1, Palmetto; 2, Cinnamon Leaf; 3, Cardium; 4, Insect; 5, Nummulite Shell; 6, 7, Fresh-Water Shells.

QUATERNARY FOSSILS.

8, Mammoth Elephant's Tooth; 9, Mastodon's Tooth; 10, Flint Implement; 11, Stone Grooved by Glacier.

insects and remains of mammals. In the Marine Tertiary are sea-shells.

QUATERNARY.

In the Quaternary drift, amongst the pebbles, sands and "wash" characteristic of gold placer beds, an occasional tooth, tusk, or bone of the great hairy Mammoth elephant or the Mastodon elephant may be discovered, together with the stone implements or bones of prehistoric man and pebbles grooved by glaciers.

CHAPTER IV.

THE PROSPECTOR'S LITHOLOGY OR STUDY OF ROCKS.

A prospector wants to know a great deal about rocks. They are his constant companions in the field. His busi-

PLATE XV.—CHARACTERISTIC APPEARANCE OF ROCKS IN THE FIELD. *a*, Sedimentary Rocks Tipped up Against Intrusive Porphyry Laccolite; *B*, *C*, Basalt Capping Soft Shales; *D*, *D*, Distant Granite.

ness is amongst rocks. He wants to be able to recognize them at sight, when he picks up a loose pebble, or confronts

48

a mighty cliff. When travelling over the mountains, as he surveys the grand panorama from the top, he wants by the peculiar forms and patterns each variety of rock is apt to take as the result of erosion and weathering owing to different degrees of hardness, to be able to make a shrewd guess from a long distance, as to whether one mountain is made of granite, or another of limestone, or a third of porphyry. This habit of forming rough guesses as to the character of distant rocks, decides him as to choosing his course for prospecting. "In those sharp granite looking peaks" he says, "may be I will find fissure veins. Yonder cones, like the spires and minarets of a Gothic cathedral must be porphyry or igneous rock, another likely locality, and mark where they break through the sedimentary strata, and tip them up all around them; at the junction of these sedimentaries with the igneous rock there may be lime-stone, and a 'contact blanket deposit.' Yon smooth grassy slopes are probably underlaid by sandstone or limestone, and the rolling valley beneath by soft shales. The latter are unpromising for precious ores." Or, again descending from his perch into the canyon below, he recognizes the granite basis, and on top of it, a series of sedimentary rocks. The lowest of these, by its rusty-white, masonry-like structure, he judges to be Cambrian quartzite, the thin-bedded strata above, Silurian limestones, and

PLATE XVI.—SAN JUAN VOLCANIC PLATEAU MOUNTAINS. Composed of a Succession of Lava Sheets Resting on Granite.

the heavy massive beds above these, Lower-Carboniferous blue-limestone, whilst a dark greenish-gray rock, running in and out irregularly among the strata, sometimes between

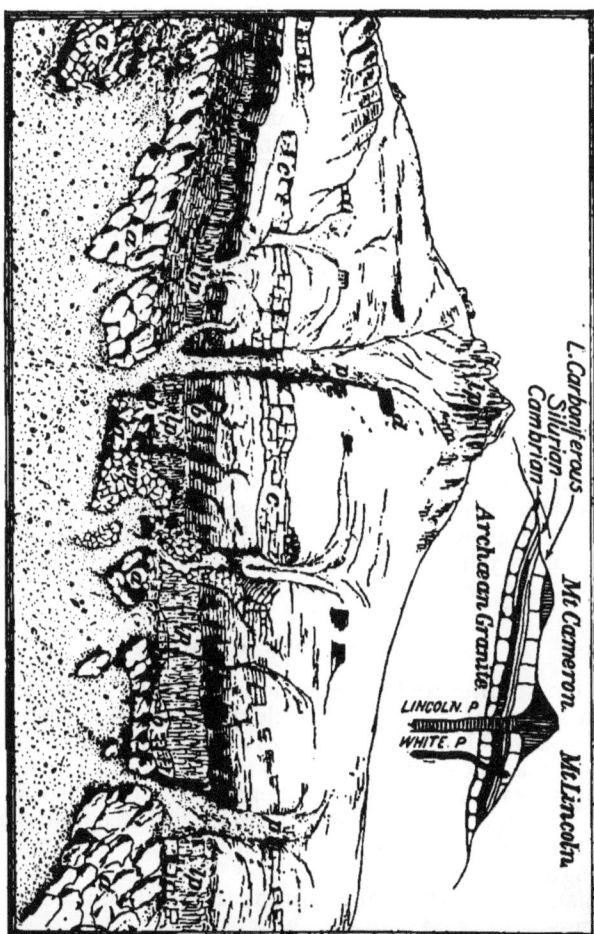

PLATE XVII.—SUMMIT OF MT. LINCOLN AND NORTH WALL OF CAMERON AMPHITHEATRE. (AND SECTION.)

a, Archæan Granite; b, Cambrian Quartzite; c, Silurian Limestone; d, Lower Carboniferous Limestone; l p, Lincoln Porphyry; w p, White Porphyry.

the stratification planes, at others cutting across them, he judges to be an intrusive sheet of porphyry, and looks again for "contact deposits." A rock running up like a low wall from the bottom of the canyon to the top, may be either a quartz fissure vein, or a porphyry dyke, and well worth examining. There are many ways of studying rocks, one by hand specimens, finding out all the minerals composing them, and then naming the rocks from which they came; another by observing the appearance of large masses of rocks in the field, and noting their mode of occurrence; and lastly if we wish to be very accurate, making thin microscopic sections and a chemical analysis, but for the average prospector these last will be rarely necessary.

If a prospector bought a manual to study rocks, for practical purposes, he would find himself amongst a sea of names of varieties of rocks, nine-tenth of which it is safe to say he would never meet with in his field experience.

To save him the trouble of wading through such books, we select just about as much as a prospector is liable to meet with in the field or find practically useful, saying little also about such common rocks as are familiar to every one.

Those that need most definition and are of most importance in the mining field, are the crystalline rocks, belonging to the class called metamorphic and igneous; the last especially needs careful determination.

Nearly all sedimentary rocks (limestone excepted) are derived from fragments of igneous and metamorphic rocks. Probably nine-tenths of the sedimentary rocks are derived from granite alone, the remainder from the igneous rocks, such as porphyry, basalt, etc. By describing the parent rock, the derivative one is more easily made out.

ROCK MAKING MINERALS.

Crystalline rocks are made up of certain distinct minerals, most of them of quartz, feldspar and mica with sometimes also hornblende and augite. Other minerals may locally occur as occasional elements.

QUARTZ scarcely needs description being so well known. The hexagonal prism of this crystal is too hard to be scratched with a knife and will scratch glass. This distinguishes it from calcspar and barite, for which it might be mistaken in the field, moreover it will not effervesce with acids.

THE FELDSPARS are nearly as hard as quartz. Their colors are white, greyish and flesh-color. They are rarely as transparent as quartz, being generally opaque. Their form of crystallization is different from quartz, and in a vein they show one smooth face of their crystal, whilst the quartz is more like crushed loaf-sugar. In a porphyry the feldspar crystals are very distinct, and give a characteristic spotted appearance to the rock. Two varieties of feldspar are characteristic of the crystalline rocks, one called orthoclase or common feldspar, a potash-feldspar, the other called oligoclase, a soda-lime-feldspar. The former is very characteristic of granitic rocks as well as of igneous porphyries, the latter is rather more characteristic of more recently erupted igneous rocks, such as diorite, basalt, andesite, etc.

Orthoclase is generally in large crystals, oligoclase in small. When the crystals are very small, it may take a microscopic examination to determine to which variety of feldspars they may belong. The oligoclase and plagioclase crystals in igneous rocks are commonly but little white dots.

To determine accurately, microscopic slides and chemical tests must be made, but this is scarcely within the scope of the prospector who wants to guess roughly at sight as to the name and character of a rock.

MICA, both black and white, needs no description.

HORNBLENDE differs from mica in being of a duller lustre and of a different form of crystallization as shown in the plate. The color is a greenish-black; the greenish tint is distinct, when the crystal is struck by a hammer.

AUGITE or PYROXENE is scarcely distinguishable from hornblende. In Colorado, augite is mainly confined to two kinds of rock, basalt or dolerite and andesite, both of comparatively recent volcanic origin. Hornblende and mica are common to nearly all the metamorphic and igneous rocks.

TALC amongst miners means almost any soft, sticky, or slippery, decomposed rock, but strictly, talc is a pale green, soft mineral like mica and is a silicate of magnesia. Steatite or soapstone is massive talc. Miners often wrongly call any soft clay or rock, soapstone also.

CHLORITE is another magnesian mineral, of a green and soft character. Chlorite is again a name given to almost any greenish rock of a schistose and soft decomposed character.

CALCITE is carbonate of lime crystal, the element of limestone, and is distinguished by softness and effervescing in acids.

DOLOMITE or carbonate of lime and magnesia is very like calcite and is the element of dolomitic or magnesian limestone. Dolomite effervesces with much greater difficulty than true limestone. To effervesce, the dolomite should be powdered, and the acid heated.

GYPSUM or sulphate of lime can be distinguished by its extreme softness, being scratched by the finger nail; it does not effervesce like lime.

BARITE or "heavy spar" occurs in some veins, but not as a constituent of rocks. It looks like calcspar, but is heavier and will not effervesce with acids.

FLUOR-SPAR occasionally occurs in veins, in cubes or massive. It is easily scratched with a knife; its colors are green, purple, yellow, blue or white.

GARNETS, GREEN EPIDOTE, BLACK TOURMALINE, and other minerals or gems may occur, but not as important constituents of the rocks.

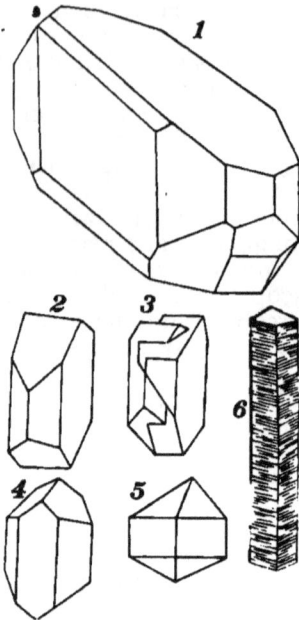

CRYSTALLINE METAMORPHIC ROCKS.

GRANITE.—Beginning with the granitic series of the Archæan age, granite proper is massive, shapeless, or amorphous and shows no bedding planes or other signs of former stratification. It is thoroughly crystalline like lump-sugar.

PLATE XVIII.

1. Triclinic Oligoclase Feldspar. 2. Monoclinic Orthoclase Feldspar. 3. Carlsbad Twins Feldspar. 4. Augite or Pyroxene, 5 and 6 Hornblende.

By some it is considered a true igneous rock, one that has been thoroughly fused by heat, as much as the lavas or molten iron; by others its crystalline amorphous condition is supposed to be the result of extreme metamorphism of originally sedimentary bedded rocks, such as gneiss or schist, the two latter being sometimes traced down through a gradual change into granite. The composition of granite is mica, quartz and feldspar with sometimes a little hornblende. The micas may be white mica (muscovite), or black mica (biotite). Both orthoclase and oligoclase feldspar may be present, but more commonly the former, which is often a pinkish flesh color. Granite, in its crystalline texture, differs both in character and appearance from porphyries and other igneous rocks, in the fact that its crystals are all jumbled up and crushed together like loaf-sugar, and none of the crystals are set like plums in a pudding, distinctly in a backing or paste of very small crystals of amorphous or glassy material, as in the porphyries or igneous rocks. Granite is probably the oldest and deepest rock known. It is often traversed by sparry veins, both great and small, which consists of quartz or feldspar or both, in a more sparry condition than when diffused through the parent rock.

These so-called "quartz veins" are often called "granulite" or "pegmatite" or "graphic granite." The quartz and feldspar are often arranged in parallel plates, giving on

GRANITE

PLATE XIX.

PEGMATITE

PLATE XX.

SYENITE

PLATE XXI.

cross-section curious marks like Hebrew characters, hence the word graphic. The bulk of our so-called quartz fissure veins in the granite mountains may be called pegmatitic veins. The colors of granite vary from reddish to gray, or nearly white to black, according to the preponderance and colors of the micas and feldspars in them.

GNEISS

PLATE XXII.

SYENITE is little more than granite in which hornblende supplies the place of mica.

GNEISS may be called "bedded granite," showing a bedded appearance. Gneiss is often curiously and prettily banded or streaked by seams of mica

CONTORTED MICA SCHIST

PLATE XXIII.

dove-tailing into each other. If mica preponderates, it is called "mica-gneiss," if hornblende " hornblendic gneiss."

SCHIST may be called laminated-gneiss or granite, being finally divided into lamina or leaves. This foliated structure is due to the arrangement of the flat-lying crystals of mica or hornblende largely composing it. It may be a mica-schist or a hornblende-schist.

SLATE is shale altered by heat into a hard crystalline structure.

QUARTZITE was originally a sandstone composed of quartz grains, which by heat have been partially fused together at the edges, resembling granules of tapioca in a tapioca pudding. Quartzite differs from quartz in being a rock made out of pieces of quartz, and not the original mineral itself. Quartzite may be white like sugar, grey, brown, or rusty. It shows a true stratified structure.

MARBLE is limestone similarly changed to a more crystalline condition.

SERPENTINE is a green magnesian rock, sometimes found with marble and igneous rocks and is formed by alteration of certain minerals in the latter.

CRYSTALLINE IGNEOUS OR ERUPTIVE ROCKS.

These are rocks which are supposed to have been thoroughly fused or melted in the bowels of the earth. Some reach the surface by fissures or volcanic vents, others have never attained to the surface or overflown it, but have intruded themselves between the weak places in the underlying strata, or have collected and cooled deep down below the surface in great molten reservoirs called " laccolites" or lakes of stone. When these have been subsequently uncovered by erosion, they may present the forms of considerable mountain masses, like the Elk Mountains, and Henry Mountains and Spanish Peaks. Geologists distinguish those rocks which have poured out on the surface from craters and volcanic vents as volcanic rocks, whilst those cooling below are called Plutonic.

INTRUSIVE PLUTONIC ROCKS.

The component minerals of these intrusive Plutonic rocks, such as are commonly called porphyries, are principally quartz and feldspar, with mica or hornblende. In color these rocks are some shade of grey, green or maroon, or even white, but their most striking characteristic is a general *spotted* appearance. This arises from more or less large, distinct, perfectly formed crystals of feldspar or quartz, set in a finer grained crystalline paste or background, standing out distinctly from it. This base or background may be comparatively coarsely crystalline, finely crystalline, or so finely crystalline, that the crystals can be discovered only by a microscope, whilst the larger crystals seem set in the paste, like plums in a pudding. In the depths of a mine the porphyry is commonly much decomposed by water action or mineral solutions, and even passes into a clay or gouge. The characteristic spotty appearance, from the presence of individual crystals of feldspar may even then identify the rock, or by chemical analysis the very aluminous character of the decomposed rock may determine its character. When feldspar is the main constituent, it is called a felsite porphyry, when a certain amount of quartz is present a quartz porphyry.

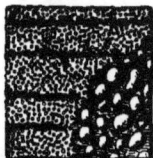

QUARTZITE (PART MAGNIFIED)
PLATE XXIV.

DIORITE, whose crystals are sometimes porphyritic in character, hence called porphyritic diorite or porphyrite, belongs also to this intrusive or Plutonian class, differing only from the others in the fact that its feldspar is of the triclinic plagioclase kind rather than orthoclase. Hornblende is a prominent constituent of this rock, and gives it, more or less, its dark, olive green tint. In appearance it resembles a dark syenite, but its occurrence as an eruptive, intrusive rock distinguishes it, as syenite is generally a metamorphic rock. The peaks of the Elk Mountains are, many of them, of diorite. Diorite or porphyrite is the so-called porphyry of Aspen, above the ore deposits.

PORPHYRITIC DIORITE (PORPHYRITE)
PLATE XXV.

QUARTZ PORPHYRIES.

These are the commonest, and may be said to be the prevailing eruptive rocks associated with our ore deposits in Colorado, as for instance at Leadville, felsite porphyries as well as quartz porphyries occur in the granite rocks in the Central and Georgetown mining districts. All these rocks are common through the West, and quartz porphyries are the most common eruptive rocks the prospector is likely to meet with in his search for ore deposits. We will describe in detail one or two typical species, though it must be observed that these porphyries are of endless varieties and shades of appearance.

FELSITE PORPHYRY
PLATE XXVI.

QUARTZ PORPHYRY.—A quartz porphyry is a porphyry that contains quartz crystals, large or small, in addition usually to large orthoclase feldspar crystals, generally of a vitreous glassy variety called "sanidin," together with small crystals of hornblende or mica. As a typical example we take that which forms the dyke composing the peak of Mt. Lincoln, Colorado, called Mt. Lincoln quartz porphyry. This porphyry and varieties of it are common in the western mining sections of Colorado.

In appearance it is a gray rock spotted with large and small crystals of orthoclase sanidin feldspar, which sometimes show an oblong face two inches long, by an inch wide, at other times a shape like the gable end of a house, according to whichever part of the crystal happens to be exposed. Sometimes two crystals are seen locked together, forming what are called Carlsbad twins. When the rock is decomposed, these crystals not unfrequently drop out and lie as pebbles on the ground. With these may be also seen rounded ends of bluish crystals like broken glass. These are portions of perfect quartz crystals, which when extracted show a six-sided pyramid at either end. These larger crystals are set in a crystalline ground mass of much smaller crystals of the same kind, together with many little black cubes of shining mica, or duller lustred, longer, rectangular, oblong crystals of hornblende. This porphyry is eruptive and intrusive, occurring in dykes, intrusive sheets and laccolites.

M^{T.}LINCOLN QUARTZ PORPHYRY

PLATE XXVII.

LEADVILLE WHITE QUARTZ PORPHYRY

PLATE XXVIII.

LEADVILLE WHITE PORPHYRY.— At Leadville there is a quartz porphyry known as the Leadville white porphyry or "block porphyry" by the miners, which needs description as it is the one that more especially is associated with the rich ore deposits. It is a white, compact, homogeneous looking rock, not unlike a shaly white sandstone or quartzite. It consists of feldspar, quartz and a little mica. Its porphyritic or

HORNBLENDE,AUGITE,MICA, MAGNETITE,GARNET

PLATE XXIX.

spotted character is so indistinct that one would be inclined to call it a felsite at sight rather than a true porphyry, but the microscope reveals perfect double pyramids of quartz and individual crystals of feldspar set in a paste of the same minerals. It is often stained by concentric rings of iron oxide and marked with wonderful imitations of trees. The latter have earned for it the title of "photographic rock" or "dendritic porphyry." These markings are only the crystallization forms of oxide of iron or manganese, something like fern-frost on a window-pane. The porphyry is very shaly, and breaks up in thin slabs; hence called also "block porphyry." It is common at Leadville and is also found elsewhere. In the same region there are many other varieties of quartz porphyry such as the "gray porphyry," the Sacramento, and the pyritiferous porphyry. The latter is often gold bearing.

YOUNGER EFFUSIVE VOLCANIC ROCKS.

These intrusive plutonic porphyries and diorites are generally older than the other class which reached the surface and poured over it and which may be called for distinction "*effusive*" volcanic rocks.

Typical of these we may cite the dark basalts and dolerites that often cap the table lands of the prairie region and overlie our coal beds. A pinkish or dove-colored rhyolite also caps some of the mesas and in certain districts an andesite lava.

DOLERITE AND BASALT.—The latter being scarcely more than a fine grained variety of the former, are very dark rocks, consisting of dark, heavy minerals, such as augite, magnetite and a plagioclase feldspar called labradorite. Such minerals are said to be basic, and the rock composing them also basic.

RHYOLITE LAVA

PLATE XXX.

ANDESITE is very like dolerite, though generally a lighter gray or pink. Both augite and hornblende may occur in it, more especially hornblende, sometimes mica also. The feldspar is called andesite feldspar from the Andes Mountains.

RHYOLITE, under the microscope, shows a peculiar flowing structure, hence its name from "rheo" to flow. The lighter rocks in Colorado and the West are generally rhyo-

lites rather than true trachytes. Their colors are pale gray, white, pink or sometimes dark.

Rhyolite consists of a fluent, vitreous, ground mass or paste, usually containing crystals of sanidin feldspar, or even of quartz. When these crystals are conspicuous so as to give the rocks a porphyritic appearance it is called "liparite."

In some cases it may have even a granite-like appearance, the crystals of quartz, mica and feldspar being more or less intermixed; then it is called Nevadite. It is an acidic rock consisting of acid minerals mainly.

TRACHYTE, from "trachus" rough, is a light colored rock, with a peculiar characteristic rough feel, due to microscopic vesicularity. It consists of a ground mass of sanidin feld-

AMYGDALOIDAL SCORIA

PLATE XXXI.

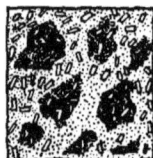

ANDESITIC BRECCIA

PLATE XXXII.

spar and augite, containing crystals of the latter. In ninety-nine cases out of a hundred in Colorado at least, also in the West, rocks which are popularly called "trachytes" are rhyolites or porphyries.

BASALTS and some of the other extrusive volcanic rocks assume a columnar form on cooling. Also, on the surface of the flow, the lava becomes minutely honey-combed like sponge, from escape of steam. This is called *scoria* and when these holes are filled with almond-shaped white crystals, *amygdaloid*. At other times the rock is a *volcanic breccia ;* that is, angular blocks of lava, great or small, are cemented together by lava. This probably was caused when the lava was pouring out of the fissure slowly, some portions congealed and were broken up by the onward flow, and again involved in the molten mass without being re-melted. Enormous masses of volcanic breccia cover the San Juan region. Sometimes, by steam, the lava is blown into dust and descending with water, is worked up into a volcanic sandstone known as volcanic "tufa" or "tuff."

OBSIDIAN is vitrified lava or volcanic glass.

CHAPTER V.

THE PROSPECTOR'S MINERALOGY.

There are two classes of minerals in which the prospector is interested, one may be called the "earthy" minerals, such as quartz, calcspar, etc., associated with the precious ores; the other, the metallic minerals constituting the ores themselves. Both of these he wants to know at sight, or to determine with the simplest appliances. Generally speaking, his eyesight, his pocket-knife, his ore-glass and a little acid, will be all he needs, nor need he concern himself about a great number of minerals, if he only knows the commoner ones well. The earthy minerals form the gangue or veinstone of the vein in which the precious ores are distributed.

EARTHY GANGUE MINERALS.

These are principally quartz, calcite, or limespar, dolomite, fluorspar and baryta, all of which we have already described among rock-forming minerals. These crystals are nearly always to be found in the adjacent rock as elements of that rock, and their more sparry condition in the gangue of the vein is derived by solution from the enclosing country rock. Thus, a vein running through granite, will contain mainly quartz, though calcite and fluorspar may be associated with it in small quantities. A vein passing through limestone naturally carries calcite or limespar. Sometimes baryta is associated with the calcite, especially if near the limestone ore deposit there are porphyries.

Baryta has been detected as an element of some porphyries which are probably ore-bearing, and when prospecting, we have found baryta to be generally an indication of ore near by, whilst calcspar, or quartz, alone, may or may not be barren. The float, or loose surface indications of ore-deposits at Aspen is commonly made up of calcspar and baryta.

FLUORSPAR in Colorado is generally confined to veins in the granitic rocks and in some of the eruptive rocks. Its presence is a good sign of ore.

OXIDES OF IRON AND MANGANESE.—These, often mixed together, form a large element in the gangue matter of a vein or ore deposit. Manganese can be recognized by its dark black color. A beautiful rose-colored carbonate-of-maganese called RHODOCROSITE is occasionally met with, associated with quartz and metal in some veins.

PLATE XXXIII.
Spathic Iron.

CARBONATE-OF-COPPER is often associated with this gangue matter. It is readily distinguished by its bright green or azure blue color. "Float" is commonly rusty with iron-oxide streaked with stains of copper-carbonate.

SPATHIC IRON OR IRON CARBONATE OR SIDERITE occurs here and there in the gangue of fissure veins. It is very like brown feldspar but heavier. These few common minerals cover nearly all that are generally met with as indications of, or in important connection with, ore deposits.

As a rule most of these minerals occur in a massive state rather than as individual crystals in a vein.

METALLIFEROUS MINERALS.

Through these gangues of various characters, the precious metals are distributed in long, narrow patches or strings, or in large crystalline masses, or in scattered crystals, or in decomposed masses. The gangue matter is generally in the majority in a vein, and the ore thinly, sparingly, and irregularly, distributed in it. When a vein is said to be ten or more feet wide, it is not to be supposed, that ten feet of solid ore is meant, but that this is the width of the gangue between walls. The ore body may be only a few inches wide. The streak or main body of ore called the "pay streak" has a tendency to keep near one wall or the other, or at times to cross from wall to wall.

HIGH AND LOW GRADE ORES.

In gold veins, flakes or wires of "free" or "native" gold occur in the decomposed gangue; and sometimes in the pure undecomposed quartz, "native" silver is found in much the same way, but more as specimens than as continuous bodies. Isolated patches of rare, or valuable minerals, such as Ruby silver, Horn silver, Silver glance, etc., occur locally in parts of the vein, sometimes coating stalactites or crystals of a "vugh" or cavity lined with quartz or other crystals. An assay from such picked specimens would give a very unfair average of a mine or prospect.

The bulk of the profits of a mine come from the commoner minerals such as galena, pyrite, or lead-carbonate, and from the average grade of the mine. In California gold mines, the average yield of gold per ton is $16. In Dakota $6. In the silver-lead mines of Leadville, $40 per ton is the average, and the ores are mostly low grade. A few mines of extraordinary high grade may yield from $75 to $100 per ton, but these are exceptional. Quantity of ore, facility for milling, cost of freight, the size of the vein, and its facility for working and nearness to market give the offset.

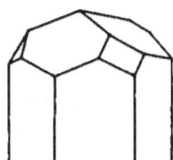

PLATE XXXIV.
Ruby Silver.

DECOMPOSED MINERALS.

Sometimes the gangue matter contains a variety of decomposed ore in rich secondary combination intimately mixed through its mass and rarely discernible by the eye. Thus yellow mud from a mine may assay high, from the presence of invisible chlorides or sulphurets of silver. No accurate estimate of the value of a mine, or even of a piece of ore, can be found, without an assay or mill-run. The reason for such richness in decomposed surface products, is, that nature has been for ages leaching out, concentrating and combining in richer forms, the essence, so to speak, of the vein.

GRAY COPPER (TETRAHEDRITE). Besides the ordinary galena and pyrites common in most mines, we sometimes find considerable bodies of gray copper in mines, or intermingled with other ores. This is generally a rich silver-bearing ore, running from 60 ounces to some thousands per ton. It generally occurs massive, rarely showing its pyramidal "tetrahedrite" crystals. In appearance it is not unlike a freshly broken piece of bronze. It is more common in fissure veins in granite and eruptive

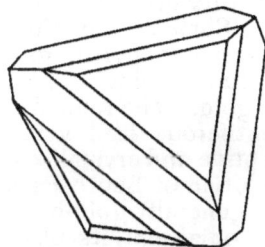

PLATE XXXV.
Gray Copper (Tertrahedrite.)

rocks than in limestone. In Halls Valley, Colorado, it is associated with baryta in a vein in the gneiss. It occurs in the Georgetown veins in granite. In the San Juan district it occurs also associated with baryta in the Bonanza mine ; and an ore not identical with it in composition, but very like it in appearance, called bismuthinite, consisting of bismuth, antimony, copper and silver, is characteristic of that region and is rich in silver. *Bismuthinite* has a more shiny tin-like appearance than gray copper, and the red color which bismuth gives to charcoal under the blowpipe readily distinguishes it from gray copper,

LOCAL VARIATIONS IN VALUE OF ORES.

There are locally in different mining districts considerable differences in the value of certain minerals and ores. In one district gray copper may rarely exceed 60 ounces of silver, in another it is invariably over 100 ounces.

A coarse galena is generally poor in silver, while fine grained "steel galena" is generally rich in silver, but the reverse may also be the case. In some of the mines at Aspen, fine grained galena, especially near the surface, is quite poor in silver, while in other mines in the same district it is exceedingly rich. Localities occur also where coarse-grained galena runs well in silver and is richer than fine-grained galena. This is the case at the Colonel Sellers mine at Leadville. So one mining district or even one mine is not a rule for another.

PYRITES.—Iron pyrites and copper pyrites, common in most of our quartz veins in granite and in the eruptive rocks, may yield both gold and silver, but usually the former. There are certain districts more characterized by pyrites than others, such as the Central City district. These are generally gold-producing districts. Some of the mines at Breckenridge and South Park have strong pyritiferous veins in eruptive dykes, such as the Jumbo mine. These have of late produced a great deal of gold. The same district, however, produces large argentiferous lead veins. Pyrites generally favor the granite, eruptive and crystallized rocks. The quartzites of the Lower Silurian of South Park and Red Cliff are often pyritiferous and generally gold-bearing. In limestone the pyrites is rare or absent, its place being filled by some form of iron oxide. In the deeper mines of Leadville, however, this iron oxide is beginning to pass down into the iron sulphide or pyrite from which it was derived. Iron pyrites can generally be distinguished from

copper pyrites by its paler, more brassy color, by its superior
hardness and by its crystallizing in cubes. Copper pyrites
is much yellower and softer, and crystallizes in a more
pyramidal form. A vein may glitter with showy pyrites and
yet be quite valueless. It usually yields more gold in its
decomposed, oxidized condition than in its unaltered state.
In the one case the gold is free-milling, and in the other it
must be smelted at much greater expense.

SULPHURETS.—This term amongst miners is loosely used,
and often means some decomposed ore whose ingredients
cannot be determined at sight, but which somehow assays
high in silver. True sulphuret or sulphide of silver is a
name embracing a large family of rich silver ores, among
which are stephanite or brittle silver, argentite or silver
glance, sylvanite or graphic tellurium, and polybasite.

All these rich ores are compounds of sulphur and silver
and other ingredients in varying proportions. They are
somewhat alike in appearance and not always so easy to distinguish.

ARGENTITE, silver glance, or sulphuret of silver, is of a
blackish, lead-gray color, easily cut with a knife, and consists of an aggregate of minute crystals. Its composition in
100 parts, is sulphur 12.9, silver 87.1. Under the blow-pipe
it gives off an odor of sulphur, and yields a globule of silver.

STEPHANITE, or "brittle" or "black" silver, is closely
allied to argentite. Its composition is
sulphur, antimony and silver, silver being
68.5 per cent. The crystals are small.
Under the blow-pipe it gives off garlic
fumes of antimony and yields a dark
globule from which, by adding soda, we
get pure silver.

POLYBASITE, common at Georgetown
and in some of the Aspen mines, such as
the Regent or J. C. Johnson, on Smuggler
Hill, is like the others, but of a more
flaky, scaly and graphitic appearance. It
is not unlike very fine-grained galena, PLATE XXXVI.
yielding 150 to 400 ounces of silver per   Stephanite.
ton.

These sulphurets sometimes line little cavities in limestones with a dark sooty substance, which under the microscope proves to be crystals of one of the sulphurets of
silver. Sometimes also a rock is stained all through a
blackish gray by these sulphurets. Iron or manganese may

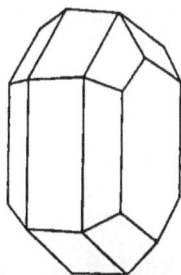

produce much the same effect, but an assay will soon reveal
the difference. Associated with such a rock we may see
flakes or wires of native silver that have emerged from the
sulphide state.

CHLORIDES.

CHLORIDE OF SILVER ("Horn silver," or Cerargyrite).—
This is another result of secondary decomposition from a
sulphide state (silver sulphide). It is a greenish or yellowish
mineral, like wax, and easily cut with a knife. It is a very
rich ore running 75.3 per cent. silver, the remainder being
chlorine. As a secondary product of decomposition it is
generally found near the surface or in cavities, sometimes
deposited on calcite or other crystals. In the mines at
Leadville it is commonly associated with other decomposed
ores, such as carbonates. In the Chrysolite mine, a mass
weighing several hundred pounds was found. Chloride,
bromide and iodide of silver are closely related, being compounds of chlorine, bromine, iodine and silver. It is notice-
able that these salts are the elements of sea water, and that
these ores are often found in marine limestones. According to Mr. Emmons, the change at Leadville from sulphide
to chloride was produced by surface waters ; these waters
are found to contain chlorine, which they probably derived
from passing through the dolomitic limestones which contain chlorine in their crystals, and these limestones perhaps
originally derived it from the sea water in which they were
deposited. Chloride of silver is found at Aspen and abundantly in the outcrop of mines in New and Old Mexico.

SULPHARSENITES.

RUBY SILVER (Pyrargyrite and Proustite).—Composed of
sulphur 17.7, antimony 22.5, silver 59.8 = 100. Crystallizes
in rhombohedrons, is seen in spots or crystals on a mass of
ore of a deep red or blackish tint. When scratched with a
knife it shows a bright or deep red color. In some mines
this very rich ore occurs only as specimens, but in others it
is present in sufficient quantity to largely influence the
value of the ore in bulk. In parts of the Granite Mountain
Mine in Montana, it constitutes the principal ore, associated,
however, with other mineral. It there occurs in large
masses and accounts for the extraordinary richness of that
celebrated mine. Proustite is much the same, only lighter
red, and consists of sulphur 19.4, arsenic 15.1, silver
65.5 = 100.

CARBONATES.

This term also embraces a large family, the commonest being carbonate of lead, (cerussite) and carbonate of copper, (malachite and azurite).

COPPER CARBONATE can never be mistaken, owing to its brilliant green and azure blue color. Copper stains are among the common surface signs of a "lead." It is generally associated also with rusty stains. Both are the surface products from copper and iron pyrites forming a vein below ground which may or may not be profitable. Copper stains are common enough in many rocks, but do not always lead to bodies of ore. In South Park the red Triassic sandstones are so stained, but yield no ore. Along our foothills there is quite a stained belt from Golden to Morrison and through Bergen Park. But few promising deposits of copper or other ores have been found, although handsome specimens of native copper have been discovered near Golden.

At the Malachite Mine on Bear Creek, near Morrison, a prospect was at one time opened showing a good deal of silicate of copper (chrysocolla) and malachite, but for some reason it has not been worked since.

COPPER in its native or uncombined state is rare in Colorado, and so far, we have as yet no true profitable mine. A great deal of copper is found associated with other ores, and is extracted by some of the smelters. Carbonate of copper is commonest in the limestone districts, as might be expected from the carbonating influence of limestone upon minerals in it, or mineral solutions passing through it. Carbonate of iron (spathic iron, or siderite), constitutes part of the gangue matter in some of our veins, and may also be found associated with coal seams generally, in the latter case in an oxidized condition.

CERUSSITE (Carbonate of lead). This is mostly found in the limestone districts such as Leadville. It is there known in two forms, one called "hard carbonates," the other "soft" or "sand carbonates." The crystals of this ore are small prisms, sometimes combined into a cross shape, of a pale grayish white, and might be taken for some form of carbonate of lime or gypsum, their weight, however, soon shows the difference. They are a secondary product of decomposition consisting of carbon dioxide and lead oxide ; as a carbonate they effervesce in nitric acid, and yield lead when heated. Cerussite is exceedingly rich in lead, carrying 75 per cent. The white lead of commerce has the same composition. In Leadville and elsewhere in Colorado it is

silver-bearing also, and though low in silver, the facility of
its treatment at the smelter makes it a very desirable ore.
As a rule it contains less silver than the unaltered galena,
but is more easily treated than the latter. The process of
change or derivation from a sulphide state (*i. e.*, from galena)
to a carbonate, is well shown sometimes in a piece of Leadville ore. A central cube of galena is surrounded by a grayish
green ring of sulphide of lead or anglesite, and outside this

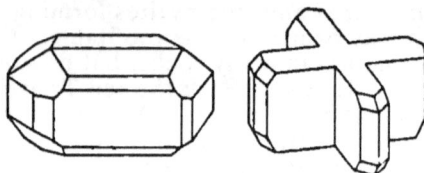

PLATE XXXVII.
Simple and Compound Crystals of Carbonate of Lead (Cerussite.)

may again occur crystals of lead carbonate. Thus the process is from a sulphide to a sulphate, then to a carbonate.
The so-called "hard carbonates" is a brown mass consisting
of a hard flinty combination of iron oxide and silica, impregnated with crystals of lead carbonate, with which are often
silver chlorides, also. The "sand carbonates" result from
the decomposition and breaking up of the hard carbonates,
or from a mass of pure crystals of carbonate of lead, which
are, by nature, loose and incoherent. The Leadville mines
are getting below these products of decomposition and
entering upon the original sulphides of galena and iron.
The yield, however, is said to be equally good.

ZINC-BLENDE (SPHALERITE), "BLACK JACK." Common in
most mines mixed with other ores. As it is a very refractory
mineral in smelting, much of it is not desirable in a mine.
It is easily recognized by its brown resinous look, or when
very black by its pearly luster. At
Georgetown, near the surface, brown
"rosin-zinc-blende" carries silver, and
is associated with rich ores, such as polybasite and gray copper. With depth the
zinc-blende becomes more abundant and
blacker, and loses much of its silver
properties. Zinc-blende may run from PLATE XXXVIII.
nothing, to twenty dollars silver, and Zinc Sulphide
rarely as high as $100 per ton. (Zinc Blende.)

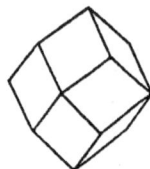

In some mines in the San Juan it occurs abundantly near

the surface and fades out with depth. We have no true
zinc mines in Colorado, the zinc being mixed with other
ores. In some mines in Pitkin County the zinc predomi-
nates over all other ores, and though it runs high in silver
the smelters do not care to take it, on account of its refractory character. In the Eastern States where zinc smelting is
a specialty, such ore might be separated and both silver and
zinc saved. In Missouri zinc and lead are found together.

In Colorado there are no mines of one mineral alone, as
in some other parts of the world. We have no true lead,
zinc or copper mines; these baser metals are either argentiferous or auriferous, and their baser qualities are sacrificed
for their richer ones.

CHAPTER VI.

ORE DEPOSITS.

THEORIES REGARDING THE ORIGIN OF ORE DEPOSITS.

A prospector will find both a practical as well as scientific
interest in considering the origin of ore deposits. Where
do the precious metals come from? What is their origin?
How are mineral veins formed and how do precious metals
get into them?

The remote origin of metals is a matter of speculation.
They may have formed part of that gaseous mist from which,
according to the nebular theory our planetary system was
evolved. As this passed into molten condition the metallic
vapors may have separated into various combinations and
consolidated and been arranged in the general make up of
the world according to their specific gravity. Some have
thought that the interior of the earth may be more metalliferous than the surface crust since the earth grows heavier
toward the center. Volcanic rocks coming up from depths
unknown contain a large per cent. of the heavier metals,
particularly iron. But we turn from these speculations to
theories of more practical interest to the prospector.

A prevalent theory amongst miners and prospectors is
what may be called "the igneous theory" or the fiery origin
of veins and metals. They are apt to attribute the fissures
themselves to some violent volcanic outburst, and consider

68

the quartz gangue or veinstone, together with the metals, as molten volcanic emanations filling at one time a wide gaping fissure.

Others demand an intense heat considering that the metals in the veins were reduced in the bowels of the earth by intense heat to a vaporous condition, which, ascending through the fissures, condensed and consolidated in a crystalline form in the upper and cooler portions of the fissures, as certain sublimed mineral vapors from a smelting furnace sometimes collect and recrystallize in the flues.

By many prospectors every indication or surface appearance of a vein, or even a likely-looking rock, is called "a blow out," a term suggestive, at least, of some sort of vol-

PLATE XXXIX.
Fold Passing into Fault Showing Broken Character of Fault Fissure and Adjacent Rocks Producing Later a Brecciated Vein and " Horses."

canic explosion at that point. With them, the "fire and brimstone" origin of ore deposits is as deep seated as the veins in the rocks.

These ideas contain a measure of truth, and were naturally suggested by observing that our ore deposits are so generally associated with volcanic rocks and evidences of past heat; and it cannot be denied but that the presence of these volcanic rocks had more or less to do with the ore deposits.

The modern study of ore deposits inclines to the belief that we need not draw directly upon the unknown profound supposed ignited regions of the earth's interior for the direct source of metals found in the veins, nor entirely from violent explosive volcanic agencies, nor from very intense heat, but rather that we may look nearer home for the immediate source of both metals and veinstone, namely, in the elements of the common country rock adjacent to the ore deposits; and for the medium of distribution and concentration of ore and veinstone from nothing more violent or volcanic than water, more or less heated and alkaline. Nor is it so absolutely necessary to suppose that the filling of a vein fissure with quartz or metal must needs come *up* from profound depths, and from a foreign source; but quite as likely from the adjacent sides of the fissure, or even from above the position later occupied by ore.

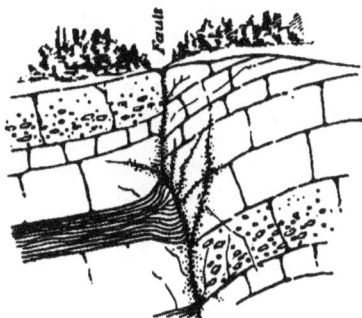

PLATE XL.

A Tight Fault Crevice Being Attacked by Solutions Producing Finally a Narrow Fissure Vein—Small Dots=Ore Solutions.

Veins of whatever kind are not vents for molten volcanic matter, but simply courses for water, more or less heated and alkaline, in fact, channels of mineral hot springs carrying earthy minerals and metals in the same solution, and depositing them, partly by cooling and sometimes by chemical precipitation and mainly by relief of pressure in such openings or weak places, as may be found convenient.

The origin of these openings and weak places in the earth's crust is various. The class of great fissures holding "fissure veins," cleaving our mountains from top to bottom to an unknown great depth, were caused by the fracturing and faulting of rocks, in the gradual process of folding upwards, and elevation of the mountain system, a process so slow and gradual that it may be even progressing now without one noticing it. The relief of

PLATE XLI.

Gash Vein Fissures in Jointed Eruptive Sheet.

extreme tension from folding results finally in faulting; though the fault fissure may extend to very great depths it was probably not violent but gradual. From time to time, the shock produced by the grinding together of the walls of a fissure in a slip or jerk of only a few inches, may have given rise to severe earthquakes on the surface.

PLATE XLII.
Joints and Bedding Planes.

A great fault fissure, too, was likely to be accompanied by minor adjacent faults and also by small incipient fissures or loose fractures of the rocks, producing parallel fissures and zones of fissure veins. Other openings, occupied now by fissure veins, may be compared to those joints common to all rocks, the result of contraction and shrinkage of the granitic or volcanic rocks from a soft, semi-plastic condition to one more solid and compact. But in no case we think were the fissures now occupied by veins 50 to 100 feet wide originally wide open chasms like that which swallowed up Korah, Dathan and Abiram in Bible history, but rather cracks fitting very tightly together by enormous lateral pressure such as we see in fault cracks of the present day not yet occupied by veinstone or gangue or metal matter. These narrow cracks were worked upon by alkaline and acid solutions and enlarged by the process, the rock gradually eaten into being replaced by gangue and metal matter, a process often further assisted by the shattered character of the rock commonly found adjacent to a great fault; this shattered cavity was sooner

PLATE XLIII.
Jointed Granite.

PLATE XLIV.
Jointed Slate.

or later eaten out, so to speak, and replaced by mineral matter. Some of the broken rock being not consumed in this way, was left, forming fragments in the vein which when small are called " breccia " and when large "horses." The great "gash" fissures such as we find occupied by so-called fissure veins in volcanic sheets such as those of the San Juan region, Colorado, appear to be due not so much to great earth movements like the last, as to openings formed by cooling and contraction of the lava, somewhat as may be observed on the cooling of iron in a slag furnace. Ore deposits of lead and

PLATE XLV.

Joints in Columnar Basalt.

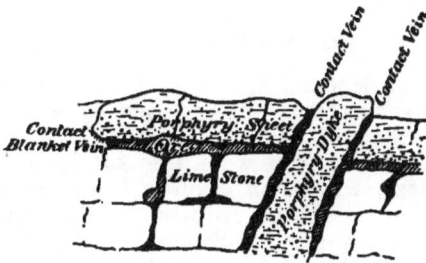

other minerals forming bedded deposits in limestones find their way in solution through the vertical joints common to all water formed rocks, resulting from contraction in consolidating from a soft, muddy condition. Such fissures are short but they act as channels to a more important line of weakness occupied by the main body of the blanket ore deposits, viz.: the dividing line between one stratum and another. Another line of weak-

PLATE XLVI.

Contact Ore Deposits Between Porphyry and Limestone.

ness for the attack of mineral solutions is at the juncture of a porphyry sheet or dyke with some other rock. The interval between them is often occupied by a " contact vein." The heat of the volcanic matter together with steam may have influenced the solutions, even if the porphyry did not actually supply the metallic element in the vein.

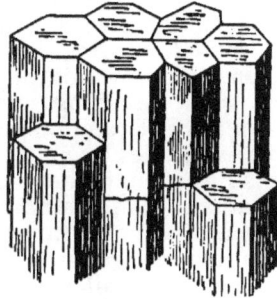

FOLDING AND FAULTING.

In the many and great upheavals of the earth's crust, resulting in continents rising above the sea, and on those

continents still greater and sharper upheavals forming
mountain ranges, rocks have been much broken and fractured, from great fractures, forming fissures miles in length
and depth, down to little cracks of but a few inches. Much
of this fracturing has been caused by the folding and
crumpling upwards of strata into mountains, accompanied
by great crushing and mashing together of the rocks.
When this lateral tangential folding and compression of the
rocks reaches its maximum intensity, the rocks break, and
a fault or slip is the result, with its attendant fault-fissure.
This relieves the strain for a while, but the shock, doubtless at the time accompanied by earthquakes on the surface,
resulted in a general breaking up of the adjacent country
into many parallel and smaller faults and cross faults, besides a general shattering of the ground intermediate to
the faults. A region thus faulted and shattered is just in
the desired condition for forming a future mineral belt or
mining region, when the cracks and scars thus made have
been healed and filled up by mineral matter, brought in
through the agency of watery solutions more or less alkaline or heated.

INTRUSIVE IGNEOUS ROCKS.

When these fault fissures descend to a very great depth,
they may tap the molten rock reservoir supposed to lie beneath great mountain ranges, and the molten lava or
porphyry, rushes upward through the weak line of the fissure, fills it with its matter, which on cooling becomes a
dyke instead of a mineral vein. These eruptive rocks may
or may not reach quite to the surface and overflow it in a
lava sheet. If they do not, they find relief by intruding
themselves laterally between the layers of stratified rocks,
whose leaves or bedding planes may have been partially
opened, like the leaves of a crumpled book by previous
action of folding. In such cases the porphyry dyke or intrusive sheet may, if it be mineralized, answer all intents
and purpose of a mineral vein, or the ore may be found on
one or both sides of such a sheet, in the line of separation
and weakness between it and the adjacent strata, or it may
permeate and mineralize by a "substitution" process an
adjacent porous or soluble rock such as limestone. Thus
both in the dyke or intrusive sheet itself as well as at its
contact with other rocks, the prospector should look for
signs of precious metal.

If the dyke or sheet should be decomposed, clayey and

rusty, it may contain free gold disseminated through it, which, at a depth which may or may not be ever reached by mining, passes into the auriferous iron-pyrites from which the free gold originally came. In this case the ore will be no longer "free" or "free-milling," but of a character that must be subjected to the more expensive treatment of roasting or smelting. Little stringers or veinlets of quartz, if observed in such an eruptive rock should be carefully examined as the most likely source of the richest gold ore. Some of our most noted gold mines in the West are in these "rotten" mineralized dykes or eruptive intrusive sheets. "Likely signs" in such would be rusty "gossan" stains of green carbonate of copper and gouge or clay matter. It is worth observing that the dyke may be only valuable as a mine as far down as the decomposition lasts and as long as the ore continues in a free state. With depth, the pyrites of the undecomposed lower portion of the dyke may be found too poor in gold to pay for smelting even.

As this desirable state of decomposition is the result mainly of the action of surface waters, a prospector may consider sometimes, where, on the outcrop of such a dyke, the rock is most likely to be deepest affected by surface action; for example, more probably below the old stream bed than on the top of a mountain, but this is not always the case. Most dykes and intrusive sheets when mineralized, are mineralized by pyrites, rather than by galena, hence they are generally more gold-bearing than silver-bearing. The contact deposits adjacent to a volcanic rock, may have been aided in their deposition by steam issuing from the molten mass, or by heated waters or steam ascending with it, or generally by the heat of the dyke, as heat together with moisture is a great solvent of rocks and promoter of chemical action.

In granitic rocks, if a "contact" deposit occurs adjacent to a porphyry dyke, it is usually a quartz vein, or a vein composed of quartz and feldspar, commonly called "pegmatite." Such contact fissure veins may be on one or both sides of a dyke. The Telluride veins of Boulder and the gold and silver veins of Idaho Springs, Central and Georgetown in Colorado are often so situated.

CONTACT DEPOSITS.

When a porphyry sheet intrudes itself into limestone as at Leadville, the ore may be looked for on either side of this sheet; but more commonly below it. At first the ore seems to permeate the limestone immediately at the line of contact, but from this somewhat horizontal line, it is apt to run down through joint cracks in the limestone, enlarging the cracks by solution, and substituting or replacing the dissolved rock with silver-lead ore, by a process called " metasomatic substitution."

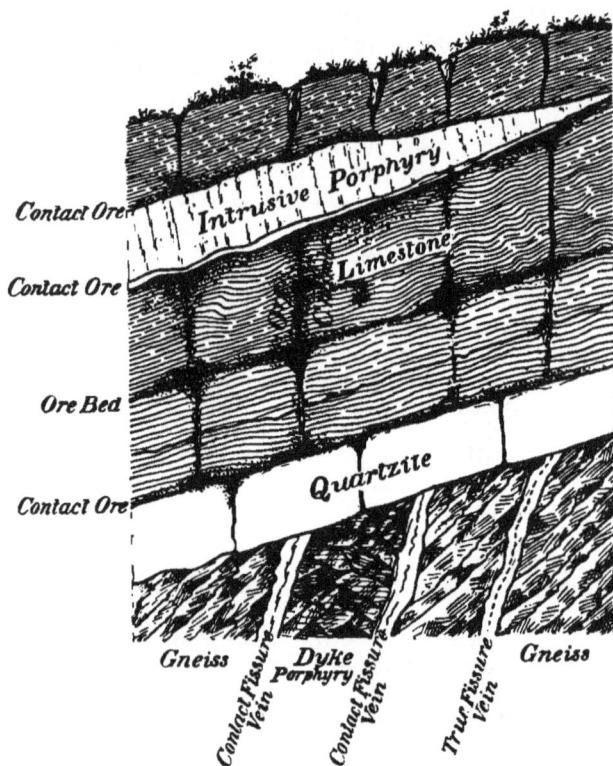

PLATE XLVII.
" Contact Blanket " Ore Deposits and " Contact Fissure Veins." ·

" Metasomatic " means literally "an interchange between one body and another." In this case it is an interchange between metal and limestone, by which the limestone is gradually replaced, molecule by molecule, with metallic matter. Thus we may suppose, that as the mineral solutions ., were working on the limestone, rotting and soaking and dissolving it, as each molecule of lime was dissolved, it was replaced or substituted by a molecule of metallic matter, until a large body of the rock was replaced by ore. This appears to be the true way in which most of our ore bodies were formed in limestone and other soluble rock, rather than that they were "washed in " and "deposited" in "pre-existing large cavities" as some have supposed.

BLANKET DEPOSITS ON BEDDING PLANES.

The solutions having worked their way down through these vertical joints, may reach a second line of weakness, viz., the bedding plane or line of stratification between one bed or stratum of rock and another, and deposit along it as on a floor. This may be between one heavy bed of lime-stone and another. If it is between two *dissimilar* rocks, such as between limestone and quartzite, or even between limestone and magnesian limestone called dolomite, it comes under the name of a "contact " deposit. Thus it is notice-able that besides great fissures, lines of weakness or "bed-ding planes" are favorite places for ore deposits, to which the natural vertical joints often act as feeders, as well as themselves containing large "pockets" or "chambers " of ore. When the deposits are confined to these "pockets " and there appears to be no "blanket " deposit, the mine is said to be "pockety," and after a "pocket " is exhausted an immense amount of money and work and blind "gophering " often follows in hunting for another pocket. There is in this case little rule to guide the prospector. Locally, by experience in the mine, he may notice that some fine line of gypsum, calcspar, or iron stain is apt to lead to a pocket and follow it. In the mines of Aspen, where the mineral zone lies irregularly but generally near about the line where the limestone becomes dolomised, a miner, when his ore "plays out," follows as closely as he can this line, which he is able to do by the different hardness of the limestone and dolomite, the latter causing his pick to "ring." In every mine there is generally some local sign to assist the miner in following up his lost ore.

SURFACE SIGNS.

The prospector in hunting on the surface outcrop for' signs of such contact or blanket or pocket deposits must

look out for signs of decomposition along the line of contact, such as lead carbonates, carbonate of copper, oxide of
iron, together with crystalline matter such as calcspar,
gypsum, or baryta. He may also observe in the vertical
joints leading down from the surface into the body of the
limestone, rusty clay fillings and iron stains. In these
" blanket," bedded deposits, prospects on a large scale may

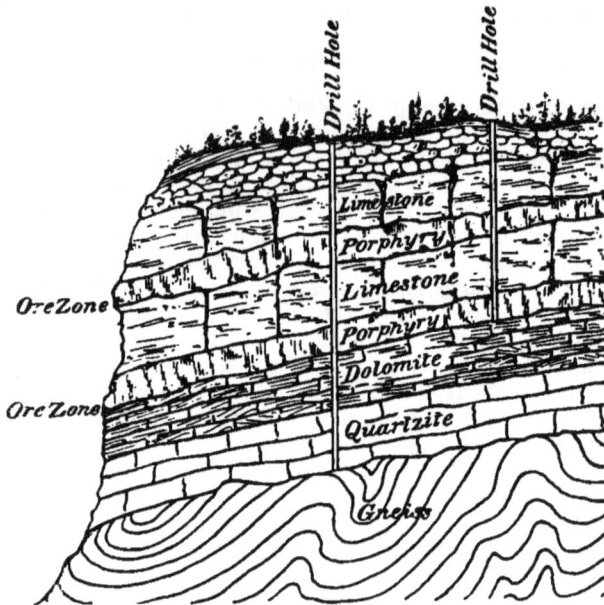

PLATE XLVIII.

Prospecting with Diamond Drills.

sometimes advantageously be done by drilling with diamond
drills from the surface down through as many of the strata
as are suspected of being ore bearing, the "cores" brought
up will show if an ore body has been penetrated together
with its approximate thickness at a certain point, and if this
process is continued over a certain area, the approximate
areal limit of the ore body may be ascertained. This work
may follow upon a close examination first of mineral signs
along the outcrop. It is sometimes done after an area has
been exploited for some time by actual mining with a view
of discovering new bodies or continuations of the ore.

Whilst profound fault cracks may be filled by lava, those not descending to such great depths doubtless lay open, till they were gradually filled by solutions carrying in earthy vein-stone and metallic matter; in a word they were the channels of mineral or hot springs. It must not be supposed that these fault cracks were ever "open chasms" commensurate in width with the wide dykes and veins now found in them, but rather in some cases very close fitting cracks, mere lines of weakness, the walls appressed closely together by prodigious lateral pressure. In other cases the fissure would be rather a shattered zone passing down through the strata, than one definite line of fissure. Doubtless when the molten lava ascended through these fissures it greatly widened them to admit of its volume. In the case of true fissure veins, the fissure or shattered zone was enlarged by the corroding, substituting power of acid mineral solutions till we have to-day a fissure vein twenty to fifty or more feet in width. In the shattered zone, this substituting process would go on easily and rapidly,

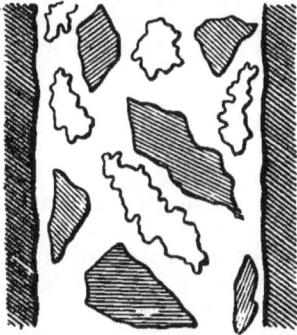

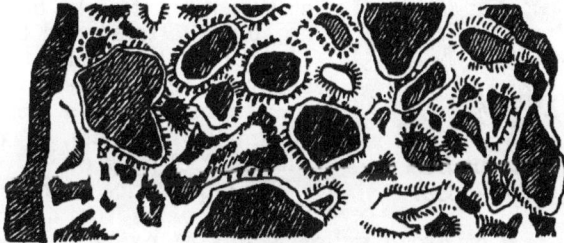

PLATE XLIX.
Brecciated Lode with Quartz Geodes.

PLATE L.
Brecciated Vein.

until nearly all the shattered fragments were replaced by mineral matter except a few "indigestible" pieces, which if

:small, would cause what is called a brecciated vein, and if large, "horses" in a vein. These fragments are not so much pieces that have fallen from above into an open fissure gradually filling up with solutions of quartz and vein matter in which they became entangled, but rather undigested, unsubstituted fragments of the wall rock, immediately adjacent to the fragments, for at times some line in the fragment corresponds to a line in the adjacent wall rock without evidence of any serious displacement. Again, the shadowy outlines of fragments can be observed partially but not entirely replaced by quartz or vein matter. Sometimes the "breccias" are surrounded by rings of quartz or metal and called "cockade ores."

HORSES.

In the San Juan region in Colorado, where we have wonderful opportunities of observing extensive sections of great fissure veins descending the faces of cliffs on either side of a canyon for two or three thousand feet, such broad veins at intervals split up into two or three arms enclosing large fragments or "horses" of the lava country rock, and again unite to form the main vein. These veins occupy a once shattered fissure, the walls of which were originally neither straight nor regular, but shattered and cracked. The vein matter insinuated itself between the shattered portions, sometimes forming a "breccia" of small fragments, at others "horses" of large ones.

PLATE LI.

Horse or Rider.

The appearance of these great San Juan veins from a little distance is that of broad yellow stains of oxide of iron contrasted with the sombre gray of the lava rocks. In some places in this region the quartz, by reason of its superior hardness, stands up above the softer lava like a low, rusty, or white wall. Again, at other localities instead of being a bold outcrop, the vein is represented by a sharp, shallow

depression forming a narrow little ravine or trench, the path'
of a rivulet and zone of abundant vegetation.   In this case
the vein was full of decomposable minerals, such as pyrite
whose oxidation decomposition products were washed out
leaving a depression in the rocks.

So, amongst some of the indications of a fissure vein to the
prospector we may note :

1st. Brown or green stains on rocks.

2d. A bold quartz vein like a wall above the country.

3d. A narrow ravine or gulch.

4th. The path of a rivulet and exuberant growth of
vegetation.

SIGNS OF FAULTING.

As these fissure veins are generally the filling of fault
cracks, and the fissures are mainly due to faulting, a prospector should be able to recognize the surface and other
signs of faulting.

Faulting as we have said, is generally the result of extreme
folding.   So, in entering a mountain region by way perhaps
of a canyon, cutting right through it on the exposed face of
the cliffs, he may observe some of these folds or arches, low
and gentle at first, but gradually, as the range is penetrated
further, increasing in sharpness, steepness and closeness ;
with this increase we may expect faults.   The presence of
the fault may be indicated by a little "sag" or depression in
the outline of the hill, or by a line of rubbish and broken
rock descending the face of the cliff, or by a zone of exuberant vegetation, or by the pathway of a little rivulet.   He
will observe a general fractured tendency of the rocks as
they approach the fault line.   By closer search he may
notice pieces of rock polished or slickensided by the movement of the walls of the fault
slipping and grinding upon
one another.   Slickenside is
a sure proof of motion having taken place in the rocks,
and is often observed on the
walls of fissure veins.   A
much faulted region is often
marked by a step-like outline, each step representing the fallen or risen side
of a fault block.   These fault lines should be carefully examined for mineral indications, especially if the fault line is

PLATE LII.
Vein *a* Faulted by Cross-Vein B.

occupied by a porphyry dyke or a vein of quartz or calcspar. Sometimes these fault lines are totally barren, both of quartz, veinstone or metalliferous matter. They may be filled up with clay, rubbish and broken rock, or the two walls may be actually welded together by pressure accompanied by a certain amount of heat; producing local metamorphic action.

Faulting too in some regions may have occurred comparatively recently, or at least after the period most marked by deposit of mineral solutions and ore deposits, in which case the fissures may be barren or at present occupied by hot or mineral springs making veins for the future. A stupendous, comparatively modern fault, runs along the west base of the Wahsatch mountains in Utah, its line is marked by a series of hot springs.

Along the face of a canyon wall the prospector may notice some peculiar stratum near the top of the cliff and its counterpart out of place near the bottom, showing that a fault has occurred, whose amount of slip he can easily estimate or measure; but when a fault of many thousands of feet occurs, a knowledge of the different geological periods involved in the slip is necessary to estimate the amount of fall. Thus if

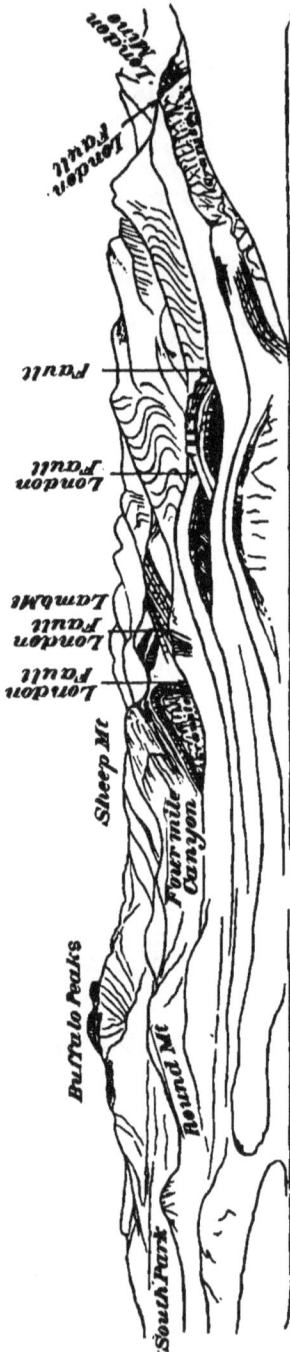

PLATE LIII. Folded and Faulted Structure of Mosquito Range, Colo.

a prospector by his geological knowledge should recognize a Cretaceous rock brought up in close juxtaposition to a Silurian rock he would know that a stupendous fault had occurred at that place, involving the entire thickness of the rocks composing the periods intervening between the Silurian and the Cretaceous.

That a faulted region is one in which great folding due to lateral tangential pressure has taken place, the folds eventually breaking down in faults, is well seen in the structure of the Mosquito Range in South Park, Colorado, which embraces the Leadville mining district.

The comparatively horizontal strata of the Park as they approach the Mosquito Range begin to fold gently, the folds gradually increasing in steepness and closeness as they approach the axis of the range. As we pass up Four Mile Canyon, which shows a complete cross section of the range, we find the axis to be formed by a magnificent and very steep arch, well shown on the face of Sheep Mountain, which having arrived at its utmost tension breaks down in what is called the London mine fault, traversing and splitting the range for twenty miles. The line of the fault is shown by a depression between Sheep and Lamb Mountain. In nearly every canyon along the flank of this range, the line of the fault is easily traced by similar arches and " sags " and by a peculiar wavy look of the turfed strata as they bend down toward the fault. As we penetrate further across the range, we pass a series of such faults, each one formerly represented by a steep fold that preceded the faulting. Hence it is that we descend from the top of this range down into Leadville and the Arkansas Valley by a series of gigantic steps or benches, each bench representing a fallen faulted block. Faults have their points of maximum depth and disturbance, from which they are apt to die out at either end in folds or rounded hills. Great faults are accompanid by minor parallel and cross faults.

The ultimate cause of this folding and faulting is attributed by some geologists to the interior of the earth growing colder and contracting, causing the surface crust to shrink and fold in adapting itself to the shrinking interior. Professor J. F. Kemp says: " The strains induced by cooling and contraction of the earth are the most important cause of fracture. The contraction develops a tangential strain which is resisted by the arch-like disposition of the crust. Where there is insufficient support, gravity causes a sagging of the material into troughs or synclinal folds which leave corresponding arches or anticlinal folds between them. Where the tangential strain is greater than the ability of the rocks to resist, they are upset and crumpled into folds from the thrust. Both kinds of folds are fruitful causes of fissuring cracks and general shattering, and every slip from yielding sends its oscillations abroad, which cause breaks along all lines of weakness."

JOINTS.

Joints, common to all rocks, appear to be due not so much to faulting and motion, as to shrinkage of the rocks in passing from a soft matter or muddy condition to one of consolidation. A good many so-called fissure veins, even in the granite series, appear to occupy extensive joint cracks, rather than fault planes. These may be due to the general shrinkage of the whole mountain mass in consolidating from a semi-plastic or aqueo-igneous state of softening to one more consolidated and rigid.

The joints in lava sheets forming curious columns like those of the Palisades of the Hudson are due to the same shrinkage from a molten state. Such joints may sometimes be mineralized for a short depth, forming what are called "gash" veins, rather than true fissure veins. The joints in sedimentary rocks are due to consolidation from a soft, muddy, incoherent condition; such joints may similarly be occupied by gash veins, or may lead to pockets or wide blanket deposits.

The line of weakness between one stratum or one set of strata and another, often a favorite line for blanket deposits, is due to one stratum being first laid down and partially consolidated before the next was laid later on top of it.

IMPREGNATIONS.

Rocks made up of loose material such as porous sandstones and conglomerates are sometimes permeated by ore solutions, as for example, the "Silver-reef" sandstone of Utah. Sandstones are frequently impregnated with iron and copper stains. In fact, if we consider that ore bodies were deposited from aqueous solutions, we have only to consider the various opportunities the rocks afford by their texture, structure, etc., for this process. Veins, in a word, are filled waterways of many and various kinds.

CHAPTER VII.

VARIOUS FORMS OF ORE-DEPOSITS.

ORE BEDS. •

"Ore beds are metalliferous deposits interstratified between sedimentary rocks of all geological ages. They lie
parallel to the planes of stratification and follow all the contortions of the enclosing strata, hence they are thrown into
folds, troughs, arches, saddles, or basins. The upper portions of the arches may often have been removed by
erosion, or the strata may be faulted." The
ore deposits or beds at Aspen occupy a faulted
synclinal fold or basin. The enclosing rock is
limestone, in part dolo-mitic. At Leadville the
deposits occupy part of a series of faulted anticlinal arches and syn-clinal troughs, of which
the Mosquito range is the main axis. The beds
lie between dolomitic

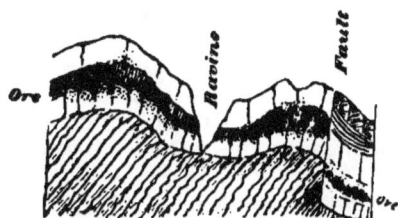

PLATE LIV.

Faulted Ore-Beds in Anticlinal and Synclinal
Folds.

limestone and sheets of porphyry. The ore beds partake of
all the folding, faulting and other contortions which the
enclosing rocks have suffered in the upheaval of the mountains.

The thickness of such deposits varies much and may
gradually thin out and disappear, but may also continue
long enough for all mining purposes.

Often there are no sharp limits between an ore bed and
the enclosing rocks, or between the ore bed and the walls,
if walls exist at all. The ore appears to impregnate the surrounding rock by a chemical interchange between the
elements of the rock and the ore. Such a "metasomatic"
interchange, "substitution," or "replacement" appears to
have taken place in the argentiferous lead deposits of Leadville and Aspen between the ore and the limestones.

According to Phillips, "a true ore bed never produces a
'combed' or 'ribbon' structure made up of symmetrical

layers, such as is common in so-called 'true fissure veins, and is usually without the crystalline texture observable in veinstones."

UNSTRATIFIED DEPOSITS, FISSURE VEINS, ETC.

Mineral veins are changeable in character, and their appearances of a perplexing and complicated nature. There is a gradual passage from one form to another, so that it is difficult to classify them. There is often no such sharp distinction between one form of ore deposit and another, as legal disputes would sometimes demand, and a witness should hardly be called upon to assert on oath that such a vein is a "true fissure," or another a "bedded vein," or a third a "segregated vein." "Nature abhors straight lines" and sharp distinctions, and delights in blending one form imperceptibly with another.

Phillips divides veins into two classes, "regular and irregular veins." "Regular unstratified deposits include true veins, segregated veins and gash veins. Irregular deposits include impregnations, fahlbands, contact and chamber deposits."

Veins are collections of mineral matter, often closely related to, but differing more or less in character from the enclosing country rock, usually in fissures formed in those rocks after the rocks had more or less consolidated.

All veins do not carry metals; some are merely barren quartz, feldspar, or calcspar, like the barren veins we so often see traversing granite or limestone rocks.

PLATE LV.

A Split Vein.

Veins may divide, "split up" or thin out, and are irregular in shape and structure, owing to the irregular width of the fissures and to other causes.

DEFINITION OF MINING TERMS.

The rock in which a vein is found is called the "country rock," *e.g.*, limestone, granite, porphyry.

The portions of country rock in direct contact with the vein are called respectively the "hanging wall," or roof, and the "foot wall" or floor. This is only in inclined or flat veins, as a vertical fissure vein can have neither roof nor floor, but only two walls, east and west, or north and south, according to the compass. The inclination of a vein to the horizon is its "dip." The horizontal direction of a vein at right angles to its dip is its "strike." The latter may commonly be observed along the surface outcrop, the former either in the workings of the mine or where the vein is exposed on the side of a canyon.

Both dip and strike of a vein often vary much, the former with depth, the latter with extension across the country. A vein or ore deposit will not unfrequently begin with a gentle dip, and increase rapidly in steepness with depth. The ore deposits on Aspen Mountain commonly begin with a dip of 25°, and at a depth of less than a thousand feet reach 60° or more.

As fissure veins commonly occupy fault fissures. their irregularities in dip and strike correspond to those we have already spoken about, under faults.

The angle of dip is usually taken from its variation from a horizontal, not a perpendicular line. Thus a dip of 75° means one that is very steep, while one of 10° is a gentle inclination.

A layer or sheet of clay called "gouge," or selvage, often lines one or both walls of a vein between the country rock and the gangue or vein proper. It is derived from the elements of the adjacent country rock, decomposed by water, and sometimes by the friction of the walls of the fissure against one another, or against the vein matter, in the process of slipping and faulting, which is often shown by its being smoothed, "slickensided," polished or grooved. Gouge often contains some rich decomposed mineral in it, such as sulphurets of silver. It sometimes occurs in the heart of a vein, especially if that vein has been re-opened anew by movements of the strata. The "Chinese Tallow" gouge of Leadville results from the decomposition of the feldspars in the adjacent white porphyry, and is a hydrous silicate of alumina.

In the granite veins in Clear Creek County the gouge is derived from the feldspars of the granite. Gouge is sometimes useful in defining the limit of the vein between walls, thus preventing unprofitable exploration into the "country." It is also a guide for following down a vein when mineral and gangue may be wanting or obscure.

Both walls are not always clearly defined by slickensided surfaces, by gouge or other mark, and so at times the vein is lost.

False walls, caused by movements in the adjacent strata, by joints, etc., also mislead.

It is not uncommon for a fissure vein to have but one clearly defined wall, the other, if it exists, being obscured or changed by mineral solutions. Sometimes two cracks or fissures occur parallel to each other and the intervening country rock has been altered and mineralized into a vein. It is probably in this way that many wide veins were formed.

Mr. Emmons has found that fissures are formed by great movements of the earth's crust or by local contraction of the rocks, and that a fissure is not necessarily one with well defined walls at considerable distances apart, filled after the formation of the fissure, but that the ordinary cracks or joints in granite quarries, extending regularly to great lengths or depths, illustrate the original fissures which have been changed by percolating waters carrying mineral solutions into veins and deposits of ore. In all crystaline and sedimentary rocks, these cracks or joints run parallel to each other at various distances apart, often plentiful and close together. In cases where percolating waters were charged with the proper metals and veinstone matter and the necessary chemical and physical conditions existed, the rocks lying between those cracks or joints were altered into ore.

PLATE LVI.

Impregnation of Rock by Vein.

As one element was dissolved another took its place, so, according to this authority, it would seem that even a fissure vein may be only a sort of "metasomatic replacement" of rock by mineral. Hence what is commonly accepted as a "wall" of a vein, is not necessarily one, and cross-cutting, in order to determine the lateral boundaries of the ore, is safer than to rely on supposed walls. A so called "slip" has often been followed by a miner as a supposed wall, until by accident he broke through and found good ore on the other side. If veins are formed according to Mr. Emmons' theory, the occasional loss of one or both walls is easily accounted for.

Cross veins of a more recent age sometimes cut or fault an older vein. The point of intersection is generally rich in mineral. Cross veins must not be confounded with "leaders," which are the filling of minor cracks extending off from the vein, and are sometimes sufficiently profitable to work. While they sometimes lead a prospector to the main vein, they may also lead a miner underground astray from the true vein.

The splitting of a vein by a "horse" or large fragment of the country lying in the vein, may be mistaken for a true cross vein, or the original fracture of the fissure may have been in the form of a star or like the spokes of a wheel radiating to the hub.

In such cases there are no true cross veins. But when, as in the San Juan district, we have two well defined sets of veins, one striking northeast by southwest, and the other northwest by southeast, they cut each other diagonally, the cut vein being the older. These opposite sets of veins have been formed at different times. Many contain a characteristically different class or variety of minerals. Thus in Cornwall, England, one set carries tin and the other lead.

SIGNS OF A TRUE FISSURE VEIN.

True fissure veins show signs of motion or slipping on the sides of the fissure, such as slickensides, gouge, crushed walls, "horses," or "breccia," the latter being small portions of the country rock surrounded and cemented by vein matter. In the Comstock, the quartz is ground to powder. The vein itself, though occupying a healed fault fissure, may be itself faulted by later movements in the mountain after the vein was formed. Some of the fissure veins on Engineer Mountain, San Juan, are so dislocated.

The vein-filled fissures being a line of weakness, may be re-opened by mountain movements, and other or different combinations of ore introduced into the heart of the vein. Such a reopening would be marked by a succession of "combs" or banded ribbon-like deposits of ore, and by gouge matter.

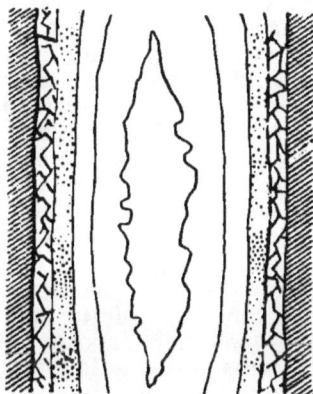

PLATE LVII.
Combed, Banded or Ribbon Structure
with Quartz Geode.

OUTCROP OF VEINS.

The outcrop of a vein is that which appears at the surface and usually attracts prospectors to the spot. Sometimes it may be, as in the San Juan district, a bold vein of hard white or rusty quartz, standing up in relief, by its superior hardness, above the surrounding country like a low wall. Or again, in the same district, from being composed of softer or more soluble substances than the prevailing eruptive lava sheets, instead of a wall it causes a depression or trough on the side of a hill, forming the pathway for a rivulet and marked by luxuriant vegetation. Commonly the outcrop consists of a decomposed mass of rock, stained with oxide of iron and streaked here and there with green or blue carbonate of copper, and is called " float " or "blossom " by the miners. This "float " is the chemically changed or oxidized portion of the true and unchanged vein lying deeper below the soil. On Aspen Mountain the float is generally a rough crystalline mass of calcspar and baryta stained with iron and copper.

In this "blossom rock" free gold is not unfrequently found, but unaltered sulphides, such as galena or iron pyrites, are rarely met with on the outcrop. In the San Juan district, on Mineral Point, we have, however, found galena at the grass roots, and broken off large chunks of it from a quartz vein outcropping on the surface.

In gold-bearing veins such an oxidized condition is desirable if it continues down to any depth, for, so far as it continues, the gold is free, and the ore is a free milling one, easily treated, and often exceedingly rich in gold, as in the celebrated Bowen mine of Del Norte; but as soon as the hard white quartz and the unoxidized pyrites of the true vein is reached, the ore is no longer free milling, but must be smelted. The gold may still be found free, perhaps, in the hard quartz, but if the pyrites should not prove rich in gold, the palmy days of the mine may be considered as past. Many such rich deposits on the surface, abounding with specimens of free gold, have proved great disappointments with depth.

WIDTH OF VEINS.

Veins may vary in width or thickness from a half inch to a hundred feet. They also pinch or widen at intervals in their downward course. The widest "mother" veins are not always the most productive, though they are very per-

89

sistent in length, and we may suppose in depth also. In the San Juan district the "mammoth" veins of quartz, often a hundred feet wide, are not the favorites for development, the ore being found too much scattered in them, and the

PLATE LVIII.
Metalliferous Veins Exposed to View near Howardsville, San Juan, Colorado.
Showing Two Systems of Fissure Veins Crossing One Another.

development less easy than in those 10, 20 or 30 feet wide, where the metal is more concentrated. These mammoth veins in the San Juan are easily traceable for miles over the surface of the country and down the sides of the deep canyons. Their limiting depth has never been reached, and probably never will be by mining.

DEFINITION OF TRUE FISSURE VEINS.

True fissure veins are popularly defined as filling fissures of indefinite length and depth, commonly occurring in parallel systems, traversing the surrounding rocks independent of their structure or stratification, and commonly, though not necessarily, at an angle different from that of the stratification—in other words, cutting across the planes of stratification. These veins originated in fissures, not necessarily wide open ones, but on the contrary, rather narrow cracks descending, however, to great depth such as those produced by faulting, or the general cleavage lines of the mountain. The latter may be frequently observed in every canyon, and also in the sedimentary rocks of the foothills and even along the flat surfaces of the plains. They are very conspicuous in the plains around Trinidad, and are there not unfrequently occupied by a series of narrow parallel dykes of basalt instead of by mineral veins. Cleavage lines or joints are familiar to every stone-quarry man.

PLATE LIX.

Fissure Vein Conforming in Part to the Bedding Planes of Stratification, in Part Crossing them.

These cracks are caused by extensive movements of the earth's crust in the process of mountain uplift, and also on a smaller scale by contraction of the rocks in cooling from a heated or molten condition, or even in consolidating from a soft or muddy condition.

The two walls enclosing a vein do not generally coincide, as might be expected, if the vein occupies a line of fault. A true fissure vein may in some part of its course coincide with the dip of the surrounding strata. As the plane of stratification or line of division between one stratum and another is a natural line of weakness, a crack once started would be liable to follow it for some distance. And when uplift occurs such places are liable to slip one upon the other, and a true parting fissure ensues conformable to the prevailing dip. Such a vein might appear at first to belong to the class of so-called "bedded veins," but if with depth it should be discovered to be cutting across the strata it would be pronounced a "true fissure vein." The appearance of slickensides or other signs of motion on the walls of the apparently "bedded portion" would then prove it to belong to the "true fissure" class, and that actual fissuring had taken place prior to the vein-filling.

CAUSE OF POCKETS IN FISSURE VEINS.

As a fault fissure in its downward course usually pursues a zigzag rather than a straight course with smooth surfaces on either side of the crack, the inequalities of one face of the crack are brought into opposition to the inequalities on the other face, as one or the other side of the fault slips up or down, and thus are produced pinches and wide cavities, which give rise to the "pinches" and "bonanza pockets" so common in fissure veins. A so-called true fissure vein may sometimes have advantages over some other forms of vein occurrence, from its persistency and comparative regularity to great depths. It must not, however, be expected that it will continue equally rich or equally poor throughout its course. There may be comparatively barren spots and rich spots, pinches and widenings, local combinations of richer or poorer varieties of mineral. But the vein as a rule is not likely to entirely give out.

RICHNESS WITH DEPTH.

There is no scientific reason why a vein should "grow in richness and size with depth." This is a popular fallacy, originating from the now less accepted theory that

PLATE LX.

Pocket and Pinches Resulting from slipping of uneven Walls of Fissure.

veins were formed by the precipitation of precious metals, by heated rising waters or vapors, and hence that the greater concentration would take place at greater depths. The "lateral secretion" theory, now by some accepted, ascribes the deposition of ore to solvent waters reaching the vein from ground quite near to it and coming naturally from above and the sides quite as often as it is ejected upward by pressure from below.

In Idaho Territory, says Mr. A. Williams, "the rule is rather that veins grow less rich and strong with depth, though strong veins may continue metalliferous to a greater depth than mining can ever reach.

"The thickness of the earth's crust which we are able to explore is very limited. Increase of heat, as in the deep Comstock mine, and other natural difficulties, limit us to a few thousand feet—3,000 at most. These deep mines have not, as a rule, proved richer with depth, but to the contrary. Some veins have been worked through alternate zones of richness and barrenness. The Comstock, which has been opened for four miles in length and to a depth of 3,000 feet, shows the ore bodies to be scattered irregularly and the barrenest ground is at the bottom. On the other hand some of the most celebrated mines derived their wealth from rich ores encountered near the surface and have proved most disappointing with depth."

Atmospheric action for a long period has often reduced the ore to its richest compound, and when the hard material is reached, leanness sets in. This, as we have observed, is commonly the case with gold veins. The richness of the Leadville mines is derived from their decomposed compounds. Again, as the surface crust can be so little explored by mining, it is to be remembered that the erosion by glaciers and waters has already removed thousands of feet of the vein, so that we are able to examine only a small fraction of it, while an unknown quantity lies in the depths below. If these veins, then, continue to the supposed great depths below, we are very far from their starting point, and erosion having removed their upper portions, we cannot find their surface finishing point; in other words, it is not a fresh "ready made" vein we find, but portions of an old vein already extensively mined by the processes of nature.

So far as our experience goes in Colorado, after a moderate depth is reached below surface action, or below the "water level," a fissure vein may grow richer or poorer, wider or narrower with depth, without any law except local experience in a district.

VEINS IN GROUPS.

Fissure veins occur in clusters and nearly parallel groups, forming a mining district, and again in that district certain peculiar veins may be grouped together, forming a "belt." Thus Boulder district occupies a certain isolated area, out-

93

side of which few mineral deposits occur for a long distance. We have also in that district several distinct belts carrying different characteristic ores, such as the telluride belt, marked by rare telluride deposits, the pyritiferous gold-bearing belt, and the argentiferous galena belt. The Central City region is characterized by auriferous pyrites belts, Georgetown district, not far distant, by argentiferous belts, and Idaho Springs, lying between the two, by both gold and silver belts.

CHAPTER VIII.

RELATION OF VEINS TO ERUPTIVE FORCES.

The ultimate cause of the richness in veins of a district or locality is, that local dynamic and eruptive forces were more energetic there than elsewhere, causing great disturbance of the rocks, accompanied by fissures, and eruptions of porphyry.

Thus at Leadville, the Mosquito range is violently folded and fractured, eruptive rocks have issued abundantly, and associated with such phenomena we find great lead and silver deposits.

Further south the great San Juan district is split up in an extraordinary manner with great fissure veins. The region is an eruptive one, consisting of prodigious flows of eruptive rocks traversed, not unfrequently, by newer eruptive dykes.

In the Gunnison district the strata have been overturned, disturbed, folded and faulted in an extraordinary manner by the intrusion of great masses of eruptive rock forming the peaks of the Elk Mountains. The strata everywhere are riddled by dykes or intrusive sheets, and the evidence of heat is apparent in the general metamorphism of the entire region. Mineral veins abound. The same phenomena are repeated more or less in the neighboring region around Aspen, and at Pitkin and Tincup.

At Boulder, Central and Georgetown there is a concentration of eruptive dykes locally in each district, and few dykes or eruptive rocks outside of those districts. On the other hand we have no ore deposits in the undisturbed rocks of the plains or the flat basins of our parks, and notably our mining districts are for the most part well into the core of the mountains, where, in the nature of things, folding. crumpling, faulting, eruptions and metamorphic heat were more energetic than along the flanks and foothills of the range which have usually proved unproductive.

The older eruptive rocks such as the quartz, porphyries and diorites of the Leadville, South Park and Gunnison districts, are more favorable to the production of ore deposits as a rule, than the more modernly erupted lavas, such as basalt or dolerite which we commonly find occurring in dykes and surface overflows, traversing or capping our Cretaceous and Tertiary coal fields along the foothills as at the Table Mountains at Golden and Trinidad.

Some of the lighter colored and somewhat recent lavas like the tufaceous rhyolite, which caps so many of the Tertiary mesas on the Divide between Denver and Colorado Springs have also hitherto proved barren. Yet the volcanic rhyolites, andesites and phonolites of Silver Cliff, Cripple Creek and Creede are productive of both gold and silver. A large portion of the eruptive rocks of the San Juan region, productive of gold and silver bearing fissure veins, are in andesitic breccias of comparatively modern date. The older eruptive rocks, as we have stated, are nearly all of an intrusive character, never having reached the surface, while the newer ones bear evidence of having flowed over the country like modern lava streams, as is shown by spongy scoria on their surface, and may be called "effusive."

In Colorado the ore body is not usually found in the heart of an eruptive sheet or dyke of porphyry, so much as at the line of its contact with some other rock, such as limestone, granite or gneiss.

CONTACT DEPOSITS.

The "contact" ore deposits of Leadville occur at the contact of quartz, porphyry and dolomitic blue limestone.

Some of the veins at Boulder, Central and Georgetown are at the contact of porphyry and granite or gneiss.

Exceptions occur, however, where mineral is found either in the heart of a dyke, or the whole dyke may be so impregnated as to constitute in a sense a vein. These exceptions are generally confined to pyritiferous gold deposits, and telluride gold deposits as at Cripple Creek.

GOLD-BEARING DYKES.

Suppose a dyke or mass of eruptive rock to be thoroughly impregnated with gold-bearing pyrites. Near the surface and often for a considerable depth the rock is decomposed and the pyrites oxidized into rusty iron ore, liberating the gold which is entangled in the "gossan" in wires, flakes or even small nuggets. As long as this decomposed or oxidized state continues, the ore is free milling, but with depth the dyke is found in its primitive hardness, studded with iron pyrites which may or may not prove rich enough for the more expensive treatment of smelting. Such gold-bearing dykes are found at Breckenridge, South Park, also in Idaho Territory, Cripple Creek, Colorado, and in old Mexico, and many other gold-bearing regions.

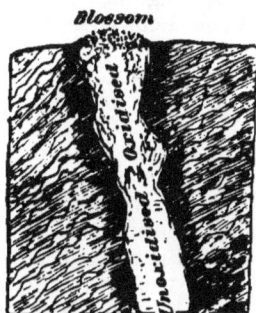

PLATE LXI.

Gold Vein or Gold Bearing Dyke, showing Oxidized and Unoxidized Portions.

The Printer Boy gold mine at Leadville is a vertical deposit in a jointing or fracture plane in a dyke of quartz-porphyry, rusty and much decomposed near the surface where it yielded free gold; with depth this passes into copper and iron pyrites. The vein is from an inch to four feet in width ; stringers carrying ore extend into the porphyry, which is highly charged with pyrites which doubtless supplied the vein with mineral through the agency of surface waters. In Arizona, near Prescott, at the Lion mine we find a green dyke of eruptive diorite penetrating granite. This dyke is traversed by numerous small veins of white quartz which near the decomposed and rusty surface are rich in free gold. At slight depth the quartz veins become charged with unoxidized iron pyrites sufficiently rich in gold to merit treatment by smelting. The surface ore is treated by a simple "arrastra," and is, of course, free milling. The gold seems to be mostly confined to the quartz veins.

FISSURE VEINS IN IGNEOUS AND GRANITIC ROCKS.

The San Juan district is an exceptional case where immense numbers of fissure veins penetrate igneous eruptive sheets. The fissure veins consist of hard gray jaspery quartz, traversing lava sheets whose united thickness is from 2,000 to 3,000 feet. The veins produce lead, bismuthinite, gray copper and other silver-bearing ores.

In Colorado true fissure veins are most characteristic of the Archæan granitic series. In fact, all the veins in that series are fissure veins. Locally they occur as in the San Juan, cutting through eruptive rocks. Outside of these formations few true fissure veins occur.

An exception may be made of the Gunnison and Elk Mountain region where the fissures traverse all the formations from Archæan granite to the top of the Cretaceous coal beds. Nearly all other mineral occurrences, such as those in the limestone regions, come under the class of bedded-veins or blanket-veins, pipe-veins or "pockets" and show none of the characteristics of slipping motion or fissure action. Under this latter class the Leadville and Aspen deposits may be grouped.

Ore deposits commonly occur at the junction or contact of two dissimilar rocks, as between quartzite and limestone or limestone and dolomite.

Lodes occur also between the stratification planes of the same class of rock, sandwiched in between two layers of limestone, and sometimes impregnating the layers on either side for some distance from the dividing line between the two strata, which is commonly the line of principal concentration of ore, and often descend from this concentration line, through the medium of cross joints, to form large pockets in the mass of the limestone. The Aspen and Leadville deposits are of this character. Also when ore bodies occupy a true fissure, *i. e.*, one cutting across the stratification planes, they may locally, for a short distance, impregnate the adjacent walls or country rock more or less. Our fissure veins in granite and gneiss often impregnate the walls to a small extent.

Mineral deposits favor as a rule the older rocks, such as the Archæan and Paleozoic series, probably because heat and metamorphic action are commoner in these older rocks which have felt all the throes of the earth from past to present times, than in the more recent ones, and such circumstances, as we have stated, are peculiarly favorable to vein formation and mineral deposition.

The bulk of our precious minerals in Colorado comes from the older Archæan and Paleozoic-series of rocks, the exception being the Gunnison region around Crested Butte, Irwin and Ruby, where ore comes from fissure veins in the Mesozoic Cretaceous rocks. The exception is accounted for by the local metamorphism, heat and eruptive phenomena of that region.

The veins in the San Juan have also been ascribed by some to the Tertiary Period, owing to their occurrence in certain supposed Tertiary lavas covering that district.

Besides heat, metamorphism, dynamical disturbances and eruptive agencies, other minor circumstances may favor ore deposition. Certain rocks, such as limestones, may offer, by their tendency to solubility and chemical reactions, more favorable conditions than others for mineral solutions to deposit by "metasomatic" interchange between mineral and limestone, until the limestone is gradually replaced by ore, much in the same way as the elements of a water-logged trunk of a tree are replaced by silica in the process of fossilization.

CHANGE OF MINERALS WITH DEPTH.

Lodes often change in the character of their minerals with depth, not only after they have left the zone of secondary decomposition and surface action, but also far below it. Thus, in the San Juan, some of the mines abound in zinc-blende near the surface, which with depth almost disappears, giving place to gray copper and other superior ores. In Cornwall, England, the shallow workings yield copper, and with depth, tin ; and locally, many such changes may characterize a particular district but cannot be formulated as a rule for other localities.

INFLUENCE OF COUNTRY ROCK.

In most mining regions, to which Colorado is no exception, a relation has been observed between varieties of "country rock" and ore deposits. Veins in passing from one country rock to another are liable to change in the size or variety of the ore, widening in connection with some rocks, and pinching or growing narrower in connection with others.

Certain rocks are notorious ore-bearers, whilst others are notoriously barren over large regions, or in special localities.

The presence of certain rocks adjacent to other different rocks has an enriching tendency on the ore bodies.

As regards rocks that are good ore-carriers or receptacles of particular classes of ore in Colorado, we may say: That quartzites and silicious rocks generally carry more pyrites, and are gold-bearing.

That veins in granitic rocks carry a greater variety of

minerals than others, and may be both gold and silver bearing.
That certain limestones carry much argentiferous galena.
That sandstones and other unaltered rocks carry little ore of any kind.

The influence of country rock on veins may be from several different causes, for instance :

Certain rocks are by their structure better adapted than others for forming regular fissures. Thus, massive limestone is better fissured than slate or shale, leaving wider open spaces for the ore to collect in.

Other rocks may be more porous, and admit mineral solutions through their pores. Of such a kind are some of our porphyries, andesites and phonolites.

Others, like limestone, are easily acted upon by solutions dissolving out the rock and replacing it with mineral by substitution.

Some are better conductors of heat, and therefore would assist chemical action and mineral solution.

And lastly, if modern theories of " lateral secretion " be true, viz.: That most ore comes from the adjacent country rock and is precipitated, substituted, or collected in the vein fissure, and further, that the metals themselves are derived from certain metallic elements in the ordinary constituent minerals of the country rock, such as mica, hornblende, or augite, it is clear that a rock composed largely of such minerals would be liable to influence the vein as an ore generator. Granite, porphyries and andesites are largely composed of these minerals.

The frequent presence of eruptive porphyry rocks near veins and ore deposits in Colorado shows that they have an important influence on those deposits, which may be of various kinds.

First, that in their component minerals and mass they actually contain the elements of the precious metals subsequently deposited in another form in the fissure vein or in the soluble limestone in contact with it.

Second, by the heat which they retain for a long time after they have congealed and hardened, they would assist in the reactions of any chemical or mineral solutions that might be on hand. Lava, at the time of its eruption, is always highly charged with steam and other gases. By reason, also, of the chemical composition of porphyry, waters passing through it would be alkaline and assist in dissolving silica and other gangue or veinstone matter, and when the porphyry has thoroughly cooled it is exceedingly porous, and being much jointed and cross-fractured, becomes like a great sponge for the absorption of all surface waters. This may be noticed at Aspen, where all the mines that are at present penetrating through the "porphyry cap" are much troubled with water, far more so than in the underlying limestone. Surface waters, then, becoming alkaline by passing through this rock, and also more or less charged with carbonic acid, chlorine, and other solvents, would be ready to dissolve both gangue and vein ingredients out of the porphyry and redeposit them in the vein fissure, or, by metasomatic substitution, in the limestone usually beneath it.

Water circulating in fissures, changes or dissolves the ingredients of the surrounding rock. The rocks enclosing lodes are always so altered, and this decomposition and alteration is not always merely local or confined to the close proximity of the ore body, but we often find a whole mining district, such as Leadville, Aspen and San Juan, pervaded by this feature. So much is this the case that it is often difficult to get a fresh, unaltered specimen of porphyry or some other country rock within the district.

The brilliant red, yellow and maroon tints that color so much of the mining district of San Juan result from the oxidation of pyrites and other iron-bearing minerals pervading the eruptive rocks, and it is noticeable that this color, resulting from alteration and decomposition, is most prominent in those parts where lodes have been discovered, as, for example, the gorgeous tints of the Red Mountain area around the celebrated "National Belle," "Yankee Girl," and Ironton mines, between Silverton and Ouray. The rocks in Geneva Gulch, Hall's Valley, Buckskin Canyon, and in other mining centers, display the same beautiful tints of oxidation in the vicinity of the mines.

"In lodes a mutual exchange takes place through the reaction of the ingredients of the rock and the materials of the vein. Thus, when water containing carbonates comes in contact with rocks or minerals containing alkalies, a chemical reaction takes place. When these last are combined with silicic acid, these silicates are decomposed by the carbonic acid and the bicarbonates. This explains both the crystallizing out of the carbonates and the so frequent decomposition of rocks containing lodes, especially those which, like our veins in granite, are feldspathic."

The same principle applies to other ores and minerals in

lodes. Thus the precious metals, in the mines of Leadville in their original condition, have been proved by depth to have been in a sulphide state, such as iron pyrites (sulphide of iron), or galena (sulphide of lead, etc.). Surface waters charged with carbonic and other acids, passing through the overlying porous alkaline porphyry and entering the underlying limestones, have, as we have previously observed, changed the sulphides into sulphates, oxides and carbonates.

The presence of a dyke near to or cutting a vein has been found often to enrich the latter at the point of contact.

In the " Colorado Central " mine at Georgetown a narrow dyke of brown obsidian traverses a large dyke of ore-bearing porphyry. The valuable ore is found close to the obsidian dyke. This might be the result of greater heat at that point. The "black dyke" in the Comstock mine is a somewhat similar case.

PREJUDICE IN FAVOR OF AND AGAINST CERTAIN ROCKS.

There is often a prejudice amongst miners in favor of certain rocks and formations, and against others. Miners who have worked perhaps in the great Comstock mine of Nevada, or the Leadville mines of Colorado, or the fissure veins in granite of the Old World, are apt to look out for and favor certain rocks and formations they find like those they have been accustomed to. Thus, as Mr. Williams says, " The peculiar 'porphyry' of the Comstock was hunted up in other districts, but did not prove metalliferous. Solid granite was looked upon by others as unfavorable, generally, because locally some granite above the gold belt of California had proved barren. Yet some of our best veins are in granite.

" Limestone was at one time a very unpopular rock and supposed only locally to produce lead, till the discoveries of Leadville, and Eureka, Nevada, overturned the scale in its favor."

In the Leadville "excitement " not only was the particular Carboniferous limestone of Leadville hunted for and prospected, but every other limestone in the South Park region, no matter what its geological age or position, was extensively prospected without results, miners not recognizing the fact that it was not limestone generally that produces rich ores, but a *particular* limestone of a particular geological period (the Lower Carboniferous) not over 200 feet thick, that happened locally to be rich near Leadville, and the reason of its being locally rich at that point was owing to the concentration of eruptive energy at that point and the intrusion of an unusual amount of porphyries, which in point of fact are far more responsible for the ore than the limestone, which happens to be merely the receptacle.

It was also quite common after the Leadville excitement to find shafts in all sorts of improbable and hopeless localities whose owners would tell you: " At Leadville it didn't matter where a man 'went down.' It was all luck whether you 'struck it' or not, and so they might as well 'go down' where they were as elsewhere." It was often said "that Leadville had exploded all so-called scientific theories about ore being in one formation or locality more than another. It was all a case of luck."

The excuse for this is to be found in the fact that in the immediate vicinity of Leadville it did scarcely matter "where you went down," seeing that that area was practically under-laid by bedded sheets of mineral, but that such would be the case elsewhere and everywhere or anywhere, experience unfortunately has shown to be untrue. It is not a particular rock or formation, but a combination of favorable circumstances that alone can make a rich mining district.

As experience advances, geologists and miners have proved that ore deposits have a much wider range than was once supposed. Formerly only the Archæan granite series was supposed capable of bearing ore deposits, because in the Old World, tin, copper and lead came principally from fissure veins in those rocks. Then deposits were found in the Paleozoic series and supposed to ascend no higher. But in the present day, and even in Colorado, they are traceable even to the Tertiary.

It is not the rock, nor the age, but a combination of circumstances, principally heat and metamorphism, that may make any rock of any period an ore-bearing one. And in prospecting in new regions it is these combinations rather than any particular rock that should be looked for.

STRIKE AND DIP OF VEINS.

The dip of veins approaches more nearly the vertical than the horizontal, usually from 75° to verticality. Nearly all our ore deposits, in Colorado, even those of the bedded class, dip more or less steeply from 25° to 75°.

For a few feet from the surface, on the steep slope of a

mountain, it is common to find an ore deposit dipping quite gently or even folded over and dipping in a contrary direction to that which it assumes with depth. This appears to arise from the weight of the strata above it tending to bend it over downward in the direction of the slope of the hill.

There is generally a prevailing dip and strike amongst a number of parallel fissure veins of a district. In the San Juan, the bulk of the fissure veins have a prevailing northeasterly strike and dip to the southeast. The angle of dip is generally between 60° and verticality.

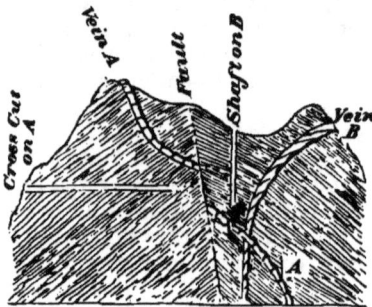

CROSS-CUTTING UNCERTAIN.

The dip as we have said, not unfrequently changes considerably with depth, usually becoming more and more vertical. From the degree of uncertainty as to the continuity of the dip, it is not always safe, on the discovery of an outcrop, to endeavor to cut it at a much lower point, so as to get the coveted depth, and better opportunities for stoping, drainage and other developments of the mine. Owing to a change of dip or fault, perhaps, the miner may have to make a much longer cross-cut tunnel than he had calculated upon before striking the vein. Sometimes, too, he may miss the vein altogether, cutting it perhaps at some point where it is exceedingly thin or poor, so poor in fact that he passes through it without noticing it or believing it to be the same vein whose outcrop looked so promising on the surface. Cross tunnels through "dead rock" should hardly be undertaken until the vein has been proved to be a strong one for a considerable depth. As we have already shown, great depths may not after all be so desirable in even a fissure vein, as there is no certainty whatever about veins becoming richer or poorer with depth. Extensive cross-cut tunnels have seldom proved paying concerns. The greatest in the United States, the Sutro

PLATE LXII.

Showing How Cross-cut Tunnels and Shafts May Miss Veins by Change of Dip or Faulting.

tunnel, six miles in length, which tapped the Comstock fissure at a depth of 2,000 feet, did not prove a financial success, and had it tapped the fissure still lower, at 3,000 feet, it would have found the vein in the impoverished condition it is to-day. It is not uncommon for a miner to strike a rich outcrop on the top of some mountain, and on the strength of its richness induce a company to run a long cross-cut tunnel in "dead rock" half through the mountain to cut this vein, and the company's resources are nearly exhausted in so doing, while the vein itself gives no returns, owing to its being left idle. Finally, perhaps, the vein is missed, or if struck, proves far poorer than was anticipated. Of course there are exceptions where cross-cut tunnels in "dead rock" may be advisable.

If a fissure vein, as in the San Juan, should outcrop near the top of a mountain and be exposed on its dip all the way to the bottom, there may be some reason for opening a tunnel in it near the base, thereby facilitating drainage, development and exportation.

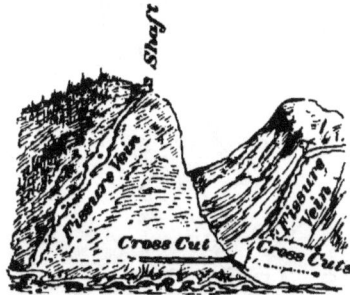

PLATE LXIII.

Fissure Vein Exposed From Outcrop to Dip.

In that case the miner is *on* the vein, with no fear of losing it; but even here, there is no guarantee that it will prove rich all the way to its outcrop a thousand feet above. "Follow your ore, and be careful how you leave it for any experimental theories," is a common and wise saying among experienced miners. We remember a tunnel in the Gunnison region which was run several hundred feet at a cost of many thousands of dollars, all through "dead rock," in the hopes of cross-cutting a certain ore body that had proved rich near the surface. At last it was given up, and subsequently a short cross-cut was made from it, and the original vein was found only a few feet from the tunnel, which had been running parallel with it all the time. The cause of the mistake was an unforeseen fault in the vein that had shifted its dip much further on one side than had been calculated upon.

CHAPTER IX.

GOLD PLACERS

PROSPECTING FOR PLACER GOLD AND GOLD VEINS.

Having given in preceding chapters a sketch of veins and ore deposits in the rocks, it follows in order to speak of gold placers, because these are derived from the former by the agencies of water, either in the form of glaciers of old, or of ancient or modern streams.

The glaciers in olden times heavily mined the rocks and the veins, by cutting broad gashes through them, thus originating the canyons. In this way millions of tons of rock were mined, together with the gold-bearing veins in them, and also the precious metals minutely diffused and scattered throughout their masses.

PLATE LXIV.

Open Placer Grounds in Canyon.

After the glaciers, the rivers took up the work, deepened the canyons, broke up the boulders and sorted them, setting free the gold and other metals they contained, and again sifted and sorted them and deposited them along their banks and in their beds.

Of the various metals thus handled by nature's jigging process, many were dissolved and destroyed by various acids in the waters, and by acids of vegetation and iron salts percolating through the placer dumps after they had been laid down. So with the exception of a few very hard minerals, such as magnetite, diamonds, garnets, rubies, etc., little remained in the placer but the imperishable gold, and even that appears to have been refined of its alloy of silver which it contained in the original vein, for placer gold is generally much purer and more valuable than that in the original vein.

In some cases, too, the fine gold disseminated through

the placer appears to have been acted upon by certain salts, such as the persalts of iron, and concentrated and amalgamated into large nuggets. Some contend, however, that these nuggets are only waterworn pebbles of gold, brought direct from the vein, the result perhaps of concentration there of the contents of large masses of gold-bearing pyrites; it is to be noted, however, that whilst gold-bearing nuggets of various sizes are to be found, not uncommonly in gold placers, they are very rarely found in gold veins.

With the gold in placers, is commonly found what is called "black sand," which is composed of grains or pebbles of magnetic iron ore, relics of the old gold-bearing pyrites chemically changed. Being near in gravity to gold, and originally associated with it, the two are generally found

PLATE LXV.
Section in Gold Placer.

together in a placer, and a prospector in surveying a bank of placer-material made up of sand, pebbles and boulders, generally looks for a streak of "black sand" as a likely place for gold. Also by reason of the gravity of gold he is inclined to look for it more down on bed-rock than in the upper looser strata.

Ancient river beds as well as those of modern rivers may be found gold-bearing, rivers that have long ceased to flow, by reason perhaps of change in the configuration of the country. In California and Australia many of these ancient gold-bearing river-beds have at a period not long distant, been deluged and covered by lava, and the gold is extracted by tunnelling beneath the lava-sheet or by shafting down through it to the gravel below. These are called deep leads whilst the ordinary uncovered gravels are called "shallow placers."

Almost anywhere along ancient or modern water courses not far from mountains, a prospector by panning, can get colors of gold even on the pebbly "wash" covering the surfaces of large portions of our plains, or even on the tops of table lands that once were plains, over which broad rivers and glaciers and large bodies of water distributed their debris, but as a rule it will only pay to work where the "wash" or "drift" or "alluvial" matter is plentiful and thick, and more than this, only where water is accessible to the work.

PROSPECTING.

A prospector hunting for a gold placer follows up the water channels in which he finds specimens of all the rocks in the neighborhood. In Australia, the prospector looks amongst these to find samples of granitic, porphyritic and quartzose rocks or clay-slate as likely signs, and also pieces of quartz honey-combed and rusty, which we have described before as "float or blossom." Plenty of broken up quartz he considers a good sign, but very pure, hard, dull white quartz is generally considered as "hungry" or "barren;" the size of the fragments denotes his nearness or otherwise to the reef, *i. e.*, the vein.

A prospector examines closely the fine sandy matter of the stream bed especially where eddies and backwater have been formed. A likely deposit should be scraped up, even down into every crevice and depression in the bed rock or solid rock bottom over which the river, modern or ancient, has worn its channel. This material should be panned. Gold, too, is often found on points and slopes of the bed rock as well as in the deepest portion. Nuggets found on high reefs above the level of the stream, imply that their weight enabled them to remain in their position, during the deeper erosion of the neighboring streams, and that the original vein from which they came, is not far off. As a rule, large nuggets and coarse gold are found much nearer to the source whence they came, than fine or "flour" gold, which is often carried to unlimited distances away out on the plains.

The character of quartz veins and of their enclosing rocks in the immediate vicinity, decides the character, too, of gravels derived from them, hence sometimes a peculiar pebble may be traced up to the peculiar rock whence it came, and the gold vein be found near it in place.

It has been observed that " leads " *following* the course or lines of a gold-bearing reef, maintain a more continuous yield than those *crossing* a number of gold reefs at intervals. Gold occurs in pockets and "shoots" at intervals, with barren portions between, which accounts for what we have stated above. In a country where the gold quartz veins are small, though rich at wide intervals, the gravels will also be small.

In very deep ground where the "wash" is very heavy a series of borings or even shafts are made to test the quality of the bank. The following points have been observed as worthy of note in prospecting for gold placers.

1. Streams crossing the lamina or stratification planes of gold reefs at right angles are likely to be richest.

PLATE LXVI.
Shallow Placer—Gold Sand in Potholes A, A and Below a Hard "Bar" B.

2. Gold is rarely found plentiful where there are indications that the current was strong, but rather in the lee under projecting points of rock, where beaches are usually formed and the water was slack.

3. Gold in streams is deposited in crevices of the "bed rock," which should be laid as dry as possible and picked up to such depths as the sand descends between the laminations.

4. Terraces are shelf-like excavations and deposits upon hill slopes above valleys, and are the remains of old glacier or river beds. The prospector should discover the inlet and outlet of the terrace and examine the gravel. The "wash" sometimes contains gold in layers one above the other.

5. Whilst working up stream attention should be paid to the banks on each side where sections are exposed so that no outcropping vein be overlooked.

6. Alluvial gold should if possible be traced to its source

whence the "float" came. When the gold is large and plentiful and the boulders large and angular the reef is likely not far distant.

7. Sometimes there is a distinct peculiar feature in all the veins of a district, such as a peculiar band of a definite color.

8. Coarse alluvial gold is not always incompatible with fine reef gold as a source, because the reef gold may be so fine in

PLATE LXVII.
Shallow Placer—Gold Sand Behind Bar on One Side of Creek.

general as to lend itself to very wide distribution when once it is liberated, while the rarer coarse grains would not be transported far.

9. Alluvial placers are richest where the current of the stream is interrupted by diminution in fall, by sudden change of direction, or by entrance of a tributary, also by reefs, bars, eddies, etc. Absolute richness depends upon local circumstances and the size and weight of floated masses.

10. Creases, holes and fissures of bed-rock over which the stream passed are favorite places.

11. The lowest layers of each separate period of deposition are the richest. *

Sometimes several different periods of deposition have succeeded each other.

12. The courses of present streams and of ancient channels are placers.

"LOAMING" is a form of prospecting. It is preliminary to such prospecting as cutting experimental trenches, or sinking trial shafts or boring. It consists in washing surface prospects from the bases and slopes of the ranges, until specks of gold, or specimens are found to be obtainable with tolerable frequency, within certain limits. The prospector then proceeds to trace the gold up hill to its source,

narrowing the limits of his work as by patient search he
approaches the vein, whence the gold has been derived.
When he can obtain surface prospects of gold up to a certain
point, or line, but no farther, he then proceeds by means of
trenching to search for the gold vein. The prospector has
often to work along a steep scrubby mountain side selecting
his prospects, numbering them, and placing samples in his
"loam bag." If he discovers prospects of gold, he finds his
way back to the spots the samples were taken from, so as to
continue his up-hill search, and trace the gold to its source
or vein. Sometimes there is no indication of a vein, soil
and bushes and debris covering its out-crop, but by loaming,
the prospector ascertains its position, so as to expose it by
a trench not many feet in length.

We remember an ingenious way in which a valuable and
long sought for vein was at last discovered. Prospectors
had long found very rich "float" at the base of a hill
whose surface was so deeply covered with loose debris
that no trace of the vein could be found. A prospector
found a small lake on top of this hill, and conceived the
idea of cutting a trench from this body of water to the
edge of the hill, and by damming up the trench, and then
suddenly letting out the water to full force, it cut a
deep trench through the loose debris down to bed rock
and the vein was discovered. This process is called "booming."

The cleavage of quartz is said to be freer, sharper and better defined, in gold-bearing quartz than in that which is barren. Pyrite is a good indication. A soft, fatty clay or gouge
often flanks the vein in its gold-bearing portions.

The mountain spurs should first receive attention for
veins; if the quartz is hard, it stands up, if soft, as it more
commonly is, it will leave a streak-like depression. On finding such, the prospector should first wash out some of the
decaying rock. If only a trace of gold is found in the quartz,
there is probably a gold vein in the neighborhood, and
trenches should be dug and exploration systematically followed up. Gold is generally near one wall of a vein, seldom
all through the stone. Quartz gold occurs in "shoots"
with barren spaces.

Before setting a valuation on a discovery, the facilities for
working the mine, such as we have alluded to, should be
considered. Placer mines as well as other mines are often
supposed to be "worked out." These are sometimes well
worth investigating and examining by cross-cuts or other

means. Sometimes it happens that more gold is obtained from "leader" veins that had been overlooked, than from the main worked vein.

Quite commonly, especially in the lower part of a placer, the pebbles and sand are firmly cemented together into a coarse conglomerate by infiltration of iron oxide and clay. This may consolidate into a false-bottom and not be true "bed rock." Generally two or three such false-bottoms, with intervening strata of greater richness, alternate with barren ones. So, many old diggings, thus supposed to have been exhausted, may be worked again, the true bottom not having been reached. These conglomerate bottoms may lie just upon bed-rock, with a white clay rich in gold beneath them. Gold occurs also in the conglomerate and must be stamped out.

Modern rivers frequently cross in their course old river courses, and redistribute their golden sands.

Placers are richer in their richer parts, than the veins from which their gold was derived.

When shallow placers are due to the wearing down of quartz veins, no placer will be found above these veins, or above the point where the vein crosses the placer. In the Sierra Nevada there is but little alluvium, the gold comes from veins near by.

Gold placers may sometimes occur below silver mines. Thus the Comstock vein was discovered by following up placer gold to its source. This vein has produced a gold-bearing silver-ore, the silver rapidly disappearing and leaving the gold behind.

EXAMPLE OF A PLACER.

In Ballarat, Australia, the "wash-dirt" runs in a series of "leads" of varying width, starting from the same point, and trending in different directions towards the "deep leads." The "reef wash" is about 100 feet deep, the "pay dirt" 5 feet. The barren drift wash overlying the "pay dirt" is of black clay. The reef itself is of green slate, the bed-rock is sandstone. Gold lies sometimes on thin layers of sand or "pipe clay" on the surface of the "bed-rock," more often in crevices of the bed-rock itself, which is more or less rotten. This bed-rock is broken up for some 12 to 20 inches and the gold is found in "pot-holes" in it 15 to 18 inches in diameter and 6 to 10 inches deep, cut out of the solid rock. The alluvial gold is found chiefly in bed-rock of slate, dipping 90 degrees. Some of these slates are soft and rotten, others are indurated. On the soft rock only is the gold found. Nuggets are found in the soft clay lying on " bedrock." Slate forms natural " riffles " for catching the gold.

Deep pools under waterfalls in gold-bearing streams rarely carry much gold. So in rivers, gold is found in "bars" or points rather than in deep pools or bends.

CHAPTER X.

"DEEP LEADS."

A "deep lead" lies deep below the surface, often covered by beds of lava, especially in California. These lava beds may be many in number, and hundreds of feet in thickness. The "deep lead" is an ancient river bed.

In the Sierra Nevada the gold is derived from metamorphic crystalline rocks of the range, partly from quartz veins in the slates, and partly from gold distributed in minute quantities all through the metamorphic rocks. The quartz veins lie between the planes of stratification of the slates, also in irregular bunches and lenticular masses of limited extent. In many localities, the rocks are penetrated in every direction by little irregular quartz veinlets, which often carry gold, and in spots are extremely rich, even where the quartz vein is only an inch thick. In some California districts, wherever a basalt capping exists, the drift beneath it is auriferous.

In California the modes of occurrence of auriferous gravel deposits are various.

"Sometimes they exist in well-defined ancient river-beds under a capping of basalt which has filled the channels of the rivers in past ages. Again, they appear in isolated mounds or hillocks, evidently the remains of such channels, which, being unprotected by a covering of lava, have been broken up by the action of the elements, also in basins or flats which have received the wash of these disintegrating rivers, also in low, rolling hills near the base of the Sierras, and beyond the reach of the lava-flows." One of the most remarkable and important gold leads is that beneath Table

Mountain in Tuolumne County. " The waters percolating through these lava-flows and reaching the gravels beneath, are charged with alkali from the lava. These alkaline waters are charged with silica in solution from the same source. Hence the fossil drift-wood of these ancient rivers has all been silicified by these silicious waters. The gravels are also cemented by the same material. These percolating waters also contained iron, for iron pyrites is found in contact with the silicified woods. In this iron-cement, gold is found in rounded grains and in minute crystals, and threads deposited by a solution of sulphate of iron at the moment of the reduction of the latter to a sulphide."

PLATE LXVIII.

Deep Placer, Table Mountain, Cal.—A A, Ancient River Channel, with Gold-bearing Gravel ; B B, Sandstones and Shales with Fossil Bones and Silicified Wood.

The dead rivers of California are on the west slopes of the Sierra Nevada, from 500 to 7000 feet above sea-level. The largest and richest lead is the "Big Blue Lead" traced 65 miles and even 110 miles. It is parallel with the main divide of the Sierra Nevada. The live modern rivers run at right angles to it, cutting canyons 1,500 to 3,000 feet deep. The "Blue Lead" runs across these ridges from 200 to 1000 feet below their summit. The lead was discovered by following up surface washings. Miners found that the modern streams were richly gold-bearing up to a certain point, increasing as this point was neared but ceasing when it was passed. These parts were in the line of the different streams, and by following up indications, the lead was eventually struck on several sections and tunnelled on. The deposit is 300 feet deep, composed of gravel, boulders, clay, and sand, on strata distinguished by degrees of fineness, by the character of the rocks, and the amount of gold, also by colors, the prevailing color being a blue-gray. Gold is coarser near the bottom, and contains a greater alloy of silver. The silver in the gold in the upper strata, has been eaten out by sulphurous acid resulting from decomposition of iron pyrites. The whole deposit is like that in existing rivers, showing banks, bars, eddies, falls, rapids and riffles. There is much gold in the eddies and but little in the rapids. The space between the boulders is filled with sand and contains gold, the bed-rock is slate.

Where dead-rivers meet, the "wash" is generally rich. Where a lead becomes very narrow, dips fast, and is inclosed between steep walls, the gold will be very sparingly distributed in holes and behind ridges and will be coarse in size.

Very large and abundant boulders in gold-bearing stream beds are often a serious obstacle in getting out the gold, from the difficulty of handling them. More than one placer has been abandoned from this cause alone.

HYDRAULICS.

Placer banks are worked on a large scale by "Giant nozzles" or Hydraulics. Before commencing such work the total depth of the placer deposit should be examined and ascertained, and the richness of the strata throughout tested. Shafts should be sunk here and there to bed rock for this purpose, and topographical surveys made to ascertain what fall and head of water can be obtained, and what outlet also for the tailings, as the latter would soon choke up the work ; the ground sometimes may be too flat to dispose of the tailings by stream-power. The choking of outlets is a fertile source of abandoning placers.

Beach Mining.—" The beach sands of the Pacific and elsewhere contain minute scales of gold and sometimes platinum, together with a great deal of magnetic iron ore. Winds, tides, and surf act as natural concentrators or separators, in parting the light and useless material from the heavier. Wind drives heavy swells on the beach at high tide together with sandy matter. At ebb of tide, the surf lashes the beach and carries back light portions of the mass with the undertow, leaving some iron sand, gold and platinum, whose weight enables them to hold their place. At low water, miners go down on the beach, scrape up the iron sand, which is generally left in thin layers, stacking it back from reach of the surf, and subsequently washing out the

gold." In some beaches much of this sand contains titanif-
erous iron ore and if attempts are made to use certain processes to save the finer gold the character of the iron may
be a formidable obstacle.

EXAMPLE OF COLORADO PLACER GOLD MINES.

California gulch, the site of the present Leadville, furnished a great amount of gold in the early days till the discovery of the lead-silver deposits in place. This discovery,
also, was due to placer mining. Whilst examining the
gravel in the gulch, Mr. Wood, an intelligent prospector, was
struck by the appearance of what the miners called "heavy
rock" some of which he assayed. His specimens yielded
27 per cent. lead and 15 ounces silver to the ton. He put
prospectors to work to find the croppings of the ore deposits, and in June, 1874, the first "carbonates in place"
were found on Dome Hill. This was practically the beginning
of Leadville. It is said that upwards of 2,000,000 dollars
worth of gold was taken out of this gulch in one summer
before the mines in place were discovered or opened up.

It is noticeable that California gulch alone furnished
almost all this placer gold, whilst Iowa and Evans gulches
adjoining it on either side, and carved out of the same series
of rocks yielded little or nothing. Why should the smaller
gulch contain exceptionally rich gravels and its neighbors
be barren?

The richest portions of California gulch were found at
bends in the course of the gulch. In one place near Oro
in the narrow bed of the gulch, a gold-bearing cement was
found containing hydrated oxide of iron, below the gravel,
yielding an ounce of gold to the ton. The gulch-gold was
worth $19 per ounce whilst that from the mines in place only
$15. The Printer Boy porphyry containing actual gold veins
in place may have been the source of some of the gold in
the gravels, together with the oxide of iron resulting from
the decomposition of pyrites in the pyritiferous porphyry as
a cementing material. Also the "Weber-grit" sandstones
at the head of the gulch have been found to carry small
gold veins, and from their abrasion also gold-bearing gravels
would have been carried down the gulch. Also of late the
rich gold deposits of Breece Hill at the Ibex and Little
Johnnie mines have been found.

"It is doubtful," says Mr. Emmons, "whether in general,
all or even the greater part of the gold contained in placer

gravels is derived from the abrasion of actual gold veins. Traces of gold may be found in a very large proportion of the massive rocks which form the earth's crust. Gold veins are concentrations of this mineral in sufficient quantity to attract attention and yield a profit. But doubtless there are a ˌvast amount of smaller concentrations which may escape notice. As the rock disintegrates and is worn away by atmospheric agencies, the gold from these smaller deposits as well as from the larger is set free from its inclosing rock and subjected to the concentrating action of mountain streams.

"Placer deposits are the results of nature's vast sluicing processes. To bring them into the condition in which they may be made available by man, requires not only the gold-bearing rock, which her agencies may grind up into sand and gravel, but the sifting power of rapid streams, which may carry down the lighter and coarser material, and a suitable channel, in which the heavier particles may lodge, as in the riffles of a sluice box. All mountain gravels, all sands of rivers coming from the mountains, contain a certain amount of gold, but it is only under peculiarly favorable conditions that the gold is so concentrated as to render the gravel remunerative.

"Among the most favorable of these conditions is a comparatively narrow channel having a hard and compact bedrock, and ridges or bends in its course, which by causing a partial arrest in the rapidity of the current shall allow the heavier particles of gold to settle to the bottom, and hold them there when once they have settled.

"From this point of view there is a very evident reason why California gulch should have furnished rich placers, and why the gold which may exist in Iowa and Evans gulches should not yet have been extracted even though the detrital material which has been carried down the gulch should originally have been equally rich in gold.

"California gulch is a valley of erosion, formed entirely by the action of running water, and since the glacial period. It has therefore a bottom or bed of hard rock. Its transverse section is V shaped and therefore favorable for the concentration of heavy particles at its bottom. When comparatively full of water, its numerous bends formed eddies in the down flowing currents, and allowed a longer time at these points for the settling of the surface particles, and as it cuts across many different formations in its course, its bed must have transverse ridges, which have caught some of the gold and prevented it from being carried farther down the stream.

" Evans and Iowa gulches on the other hand are glacier-carved valleys. Their courses are straight, their bottoms broad and comparatively smooth. The glacial moraine material with which they are largely filled has not been subjected to the sifting or jigging process to which gravel is subjected in the bed of a stream. The lower part of their present beds is cut, not out of rock, but out of the loose gravelly formation of the 'Lake beds.' This later bed, along which the material brought down by post-glacial erosion has been carried, has not a sufficiently hard and permanent bed-rock to allow of the concentration of gold on its surface."

ALMA AND FAIRPLAY PLACERS, SOUTH PARK.

Along the banks of the Platte river are enormous masses of glacial morainal matter consisting of boulders and sand brought down partly and principally from Mount Lincoln and receiving contributions from side glaciers of the Mosquito range. This material forms undulating banks on either side of the river. This placer "wash," from 50 to 100 feet thick, is worked for gold principally at Alma and Fairplay.

At Alma the heavy bank of "wash" is mined by the giant nozzle. The banks are also cut back into blocks of ground, by water from a flume, which is let out at intervals along the bank above ; at each place it cuts a narrow ravine in the loose debris and at the same time makes the banks easier to be attacked by the water of the giant nozzles which rapidly undermine them. The water and sand from these streams run down into the sluices, whose bottoms are paved with discs of wood, forming "riffles" to catch the gold, whilst the lighter sand is carried onward by the stream. In their "clean up" in the stream bed, they not only wash down to bed-rock, but after hunting with their knives in every crack and crevice of the latter, they dig it up for a foot or two, and further examine it. The rock is a jointed sandstone.

Quicksilver is thrown into the sluices, to collect the finer gold which is afterwards retorted. Whilst gold is found all through this bank of "wash" from "grass roots" down to bed-rock, the greatest quantity of gold and largest nuggets are found at "bed-rock" or in its interstices.

The source of some of this gold may be a series of large, but not very productive quartz veins in granite, near Mount

Lincoln, whence the main glacier originated. It is also probable that a good deal of the gold came, as said before, from the breaking up of the various rocks in which it was disseminated, more especially the porphyries and crystalline rocks.

In the winter, owing to freezing of the water supply, the work has to be discontinued till the following spring.

CHAPTER XI.

MINING REGIONS SHOWING EXAMPLES OF ORE DEPOSITS.

FISSURE VEINS IN GRANITIC ROCKS.

Having described in previous chapters the nature of veins, ore deposits, etc., and how to prospect them, it will be of interest as well as profit to the prospector, to learn something of the mines and mining regions themselves. For this purpose we propose giving a sketch of some of the leading mining regions of Colorado and the West, as instructive illustrations and examples of what we have written in previous chapters. As we said in our advice as to the education of a prospector, the best education for him is to go to, and spend as much time as he can in, the mines and mining regions themselves.

We will take first the regions characterized by *fissure* veins. These veins are in the granitic and igneous districts of Colorado. In the granitic ranges, the mining districts of Boulder county, Gilpin and Clear Creek, are the most noted, the principal mining towns being Boulder, Jimtown, Georgetown, Central and Idaho Springs.

BOULDER MINES.

The geological features of Boulder consist in a series of ridges or hogbacks rising up from the prairie and flanking the granite mountains. These represent Mesozoic strata consisting of sandstones, limestones and shales, containing beds of coal and other economic products, but no precious metal. Volcanic action has occurred in their vicinity as

shown by a large dyke of basalt at Valmont. These hogbacks, so universally present, flanking the granite mountains,
are, in Colorado, destitute of precious ores. Inside of and
west of these is the Archæan granitic front range, consisting
of heavily bedded granite-gneiss, profusely traversed by
veins of "pegmatite" or very coarse sparry granite, consisting of white feldspar and quartz, with very little mica, and
from a few inches to 40 or 50 feet in width; with these also
occur some dykes of eruptive rock, some of it a dark black
rock like basalt, called "diabase"; others are lighter colored quartz porphyries and diorites. In the telluride belt,
whilst pegmatite veins are abundant, eruptive rocks are
scarce, but west of the telluride belt, which is more or less
confined to a special area underlying the Magnolia, Sugar
Loaf, Gold Hill and Central districts, enormous masses of
eruptive rock are found, but no tellurides. In the non-telluride districts, such as Caribou, Ward and Jimtown, rich silver
ores are found associated with galena, gray copper, etc.,
and gold ores associated with copper and iron pyrites.
Thus there are two or three distinct belts in the region,
a telluride gold belt, and a silver belt, and a gold pyrites
belt. It is noticed that the entire region has been locally
disturbed by volcanic forces, and volcanic rocks abound;
outside of this disturbed region there are no mines for a
long distance.

The Boulder mines are celebrated for the occurrence of
telluride minerals, some of the richest and rarest ores occurring in nature. These ores are confined to a belt occupying the eastern part of the district, and nearer to the hogback region of the plains than any other important ore
deposits in Colorado.

West of this belt in the Caribou district the ores are
argentiferous galena, with brittle silver. In the Ward district pyrites abound, and where it is decomposed the gold is
free. The pyrites though gold-bearing are difficult of reduction.

The pegmatite veins containing the ore stand at a high
angle and are often very wide, but the rich ores, especially the
tellurides, are concentrated in thin streaks and not very continuous bodies. The gangue or vein material is simply an
alteration of the adjacent granite, or gneissic country rock,
into a more sparry, larger crystalline form, consisting of
quartz, feldspar, and some mica. This is impregnated with
rich mineral, whose source is probably not far to find, the
metal elements being microscopically or chemically diffused

through the mineral elements composing the adjacent country rock, which is sometimes porphyry, and at others gneiss. This impregnation has taken place either along the contact of an eruptive rock with the country rock granite, or else in a pre-existing vein of pegmatite, or along some fault or jointing plane in the country rock itself which has been favorable to the concentration and precipitation of metallic minerals from their solutions. The direction of the veins is generally between Northeast and Northwest, or East and West; their dips are steep or vertical.

The quartz of the pegmatite gangue, when impregnated with telluride ore, has a pale, bluish-gray and rather greasy appearance, streaked here and there with a dull, blackish, greasy stain, upon which sometimes the true telluride minerals such as sylvanite, can be seen, generally in long thin crystals of a bright tin-like appearance. It is sometimes called graphic tellurium, because the crystals crossing one another assume the form of Hebrew characters. Sylvanite is a telluride of silver and gold. There are many varieties of telluride, some rich in silver and others in gold, and some with both combined. When a piece of gangue containing tellurium is roasted, the gold comes out in good sized globules on the surface.

Two great mother-veins, called the Maxwell and Hoosier veins, traverse the telluride district for several miles, easily traceable by their rusty color. One carries pyrites and tellurides, the other silver ore and gray copper. Gold Hill district, in the telluride belt, is traversed by the Hoosier gangue. Several veins cross the Hoosier gangue and are richer in its vicinity; in some, the ore is a telluride at the surface, but with depth passes down into gold-bearing pyrites.

The Ward district outside the telluride belt carries copper and iron pyrites bearing gold. Caribou is silver-bearing, its ores are galena, copper pyrites and zinc-blende occurring in gneiss near a dyke of eruptive diabase. The No-Name vein crosses and faults the Caribou vein. Its ores carry both silver and gold; the ores are silver glance, brittle silver, gray copper, galena, copper pyrites, with native and ruby silver. The copper pyrites carries more gold than silver.

The granitic rocks near Boulder are thrown into a series of parallel folds, one series cut diagonally by another. The telluride veins run along the slopes of these folds. The veins are in cracks and fissures coinciding with this folding, some of the main fissures being filled at once by porphyry dykes, the others more gradually by vein material. The

veins occur along, on, and near these dykes, along lines at
the junction of the more massive granite with the bedded
gneiss, along and between stratification planes of schist,
and along the joint planes of granite. The veins are due to
percolating alkaline waters dissolving metalliferous material
and veinstone from the surrounding rocks. It is noteworthy
that alkaline springs still exist in the neighborhood, as they
do also at the mining district of Idaho Springs. The veins
occur where the foldings are abrupt, and the direction of the
veins is parallel to the strike of the stratification. As a rule
the veins are not of great extent. A single vein can rarely
be traced on the surface or beneath it for more than 600
feet. Before that distance is reached, the vein spurs off
again into another.

Where veins cross at a small angle or where a spur
branches off from the main vein, accumulation and enrichment of ore takes place. There are two courses of veins,
one East and West, the other Northeast by Southwest;
the former system appears to be the older as the latter
faults it.

The ore occurs in chimneys or pockets, with â good deal
of barren ground between.

Small veins run parallel with each other for some distance, the interval filled with granite or pegmatite. Sometimes a vein pinches out entirely (contrary to the general
habit of true large fissure veins occupying great fault
fissures). The ore streak is from 1 to 20 inches wide containing more of this blue, greasy, fine grained "horn quartz"
than the country rock. Some of the veins interlace like
arteries in a human body. Minute particles of pyrites
(marcasite) often produce the dark stains we have noted on
the telluride quartz. By moistening the stone, the telluride
minerals and pyrite appear distinctly.

A TYPICAL BOULDER COUNTY MINE.

A good typical and very instructive example of a contact
fissure, gold-bearing vein is that of the Golden Age at Jimtown, north of Boulder.

"At Jimtown a quartz-diorite dyke occurs, of light color
containing much hornblende and titanic iron, running nearly
through the street of the village. The cliffs at Jimtown,
over 500 feet high, are of quartz porphyry, of white color,
consisting mainly of large crystals of quartz and feldspar,
set in a fine grained crystalline ground mass or paste.

GOLDEN AGE AND SENTINEL VEINS.

From the town, the road winds up a steep mountain composed of coarse gray granite, with occasional belts of gneiss. Here are located the Golden Age and Sentinel mines.

The Golden Age covers the outcrop of a quartz-porphyry dyke cutting through the granite. This dyke varies in width, from a few feet to about fifty. The outcrop of the main ore chute of the Golden Age extends along the "contact" on the lower side of the porphyry dyke. At a depth of 100 feet the main shaft discloses a split in the vein. The hanging wall of the vein continues into the dyke, but with porphyry hanging and footwalls, until a depth of 330 feet, where it enters the upper contact between the porphyry and granite. The dyke has been much acted upon and decomposed by vein forming agencies in the upper workings, but in the lower it is less decomposed and shows considerable pyrites. The Golden Age veins are well defined, presenting a banded or ribbon structure. They are inclosed in distinct walls with gouge or selvages, which at times show slickensides. The seams and feeders that have enriched both veins come in from the porphyry dyke.

The ore from the Golden Age contains rich and magnificent specimens of free gold. It is a free milling ore. When rich, the gangue is a hard, flinty, vitreous white quartz. The gold is seldom accompanied by pyrites. It is generally imbedded in the white quartz as bright yellow gold, in size, from coarse grains to nuggets several ounces in weight; after it reaches the lower contact between the porphyry and granite and enters the granite, there is an increase in the baser metals, such as zinc-blende, galena and pyrites, but the ore still retains its value in free gold.

Returning to the surface, the Sentinel location covers the apex of a vein, which there appears enclosed in a belt of schistose or gneissic rock.

This vein dips South at an angle of 70° and passes through the Golden Age vein on its course.

The Sentinel vein ore is entirely distinct from that of the Golden Age. It is the characteristic bluish horn quartz of the tellurium veins of Boulder County, with characteristic chalcedony quartz crystals and finely disseminated pyrites. The value is in metallic gold and such tellurium ores as petzite and sylvanite. Whilst most of the gold was deposited as native gold, a portion has evidently been rendered free by partial decomposition of the tellurides. This ore is very rich. The richest ore usually occurs in two narrow seams or streaks from a foot to ten feet apart, the intervening space being more or less mineralized country rock. It is richest when in the schistose rock, and poorest when it passes through the porphyry dyke. The crossing of the Sentinel vein through that of the Golden Age is very clearly marked; it very slightly faults the Golden Age vein.

The gold mines of Boulder County belong to two distinct periods of vein formation; to one belong the non-telluride ores, and to the other those producing tellurium. The tellurium veins appear to be the later of the two.

The ores of the Sentinel tellurium vein are lower grade where the vein passes through the porphyry dyke. This is due to the Golden Age vein being formed first, and draining the dyke of its disseminated mineral values. The Sentinel received its mineral from the schistose or gneissic rocks, and is consequently richer where enclosed in those rocks than when in the dyke.

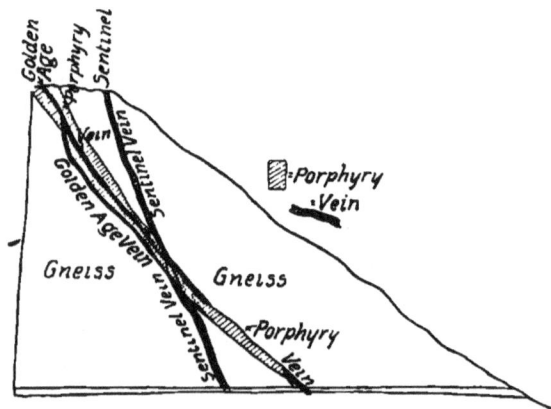

PLATE LXIX.

Section of Golden Age Vein, Jimtown, Boulder Co., Colo.

Prospectors look for richer or larger bodies of ore when veins unite or cross each other. In the Golden Age the two veins unite about 100 feet below the surface. There are similar veins of the same age, and large and rich ore bodies are found at their junction. On the other hand, the Sentinel vein of later age, passing through the earlier Golden Age vein, produced no enrichment of the ore bodies.

To form such ore bodies, the veins should be of contemporaneous origin."

The ore deposits of Gilpin and Clear Creek Counties are very similar to those of Boulder, only they do not produce tellurium ores. The country rock is the same granite-gneiss, penetrated here and there by porphyry dykes. The pegmatitic veins are either in the gneiss or between the dykes and the granite. In some cases the porphyry dyke constitutes a vein in itself, such as the Minnie, which is a felsite porphyry, and the Cyclops, a quartz porphyry, In Gilpin county, around Central City, the ores are a mixture of copper pyrite and iron pyrite with a very little galena and zinc-blende. All are gold-bearing.

The richer ore occurs in streaks not over a foot wide, in a compact, fine grained mass of pyrite. Copper pyrite is richer than iron pyrite. The rest of the vein, often many feet wide, carries pyrite irregularly disseminated through decomposed country rock. The bulk of these ores are difficult to treat, and are milled, the loss being 40 per cent. higher in the unoxidized ores than in the oxidized. The veins follow the cleavage planes of the gneiss, cutting the stratification planes at right angles with a vertical dip. The porphyry dykes are older than the veins, as the cleavage planes intersect both the porphyry and gneiss alike. For an interval of 20 miles between these mining districts and the plains, there are no ore deposits of any importance known.

In Clear Creek County the ores are mainly silver-bearing ; the silver is derived mainly from galena and gray copper. Dykes of obsidian occur in one of the mines parallel with the vein, which is itself a porphyry dyke. The richest mineral is close to the obsidian dyke.

FISSURE VEINS IN TRUE IGNEOUS ROCKS.

Whilst most of our fissure veins and ore deposits generally are more or less associated with the *presence* of igneous rocks, there are some which are essentially *in* igneous eruptive rocks alone.

The most remarkable of these are the fissure veins of the San Juan region in southwestern Colorado.

This region consists of an enormous plateau of lavas of great thickness resting upon and originally overflowing a a low mountain range or plateau of granitic and upturned sedimentary rocks, the latter representing most of the geologic periods from Cambrian to Tertiary. The thickness

PLATE LXX.—SAN JUAN VOLCANIC PLATEAU MOUNTAINS.
Composed of a Succession of Lava Sheets Resting on Granite.

of these great lava flows, which were erupted about the Eocene period of the Tertiary, is upwards of 1,500 feet; this lava mass has been cut up by glacial and river action by profound canyons, into a rugged mountain range, the summits of some of the castellated mountains reaching a height of 14,000 feet above the sea. The lava sheets are also traversed to a depth of 1,500 feet more or less, by an extraordinary number of great quartz-fissure veins. These veins appear to fill shrinkage cracks resulting from the contraction on cooling of the lava sheets, strictly speaking they are rather "gash veins" on a larger scale than "true fissure veins," for they are mostly *limited to the thickness of the lava overflows* and cease when they reach the underlying granite.

There appear to have been two principal eruptions; the first, during the early part of the Tertiary, covered the higher region of the San Juan mountains to a depth of 1,500 feet with an overflow of brecciated andesitic lava, which on cooling developed fissures of contraction traversing the lava mass in all directions; these were subsequently and slowly filled with a hard bluish quartz containing more or less ore.

Following the first grand overflow were others of less magnitude, consisting of non-brecciated andesites and rhyolites. This second dynamic movement produced locally, fissures extending below the horizon of breccia into the stratified rocks. These, however, are seldom productive below the eruptive zone. There are also metal deposits in connection with still older eruptions of andesite and diorite, such as Mineral Farm, Calliope, etc.

A, Column of Quartz and Porphyry; B, Tunnel; C, Cave in Ore Chamber; D, Bright Red and Yellow Rocks; E, Lava-Terraced Mountains.

National Belle Mine, Red Mountain, San Juan, Colo.

PLATE LXXI.

RED MOUNTAIN.

In the Red Mountain district the ore deposits form a peculiar group. They occupy a series of more or less connected irregular chambers, trending downward, probably channels of ancient hot mineral springs. The mineralizing water completely silicified the surrounding eruptive rock for some distance away from the ore chambers. So the ore bodies are distributed through a huge irregular column of quartz extending to an undetermined depth.

Large masses of brilliantly colored material are conspicuous in this region. They have been acted upon by mineral waters circulating through their crannies and fissures. Ore bodies are occasionally found in these and such mines are locally known as cave mines.

The ores of the San Juan are mostly argentiferous gray copper, copper pyrites and galena associated with zincblende and iron pyrites in usually a hard horn-quartz matrix. Some of the ore locally contains a high per centage of bismuth; others produce pyrargyrite and polybasite, rich silver minerals; others carry considerable gold, such as the recently discovered gold belt at Ouray. This belt occurs in Dakotah Cretaceous sandstone, which has been altered into a quartzite by the intrusion of dykes and sheets of eruptive diorite. One of these sheets spreads out in the quartzite. The ore occurs at the top of the quartzite, at its junction with a bed of shale. The gold, which is free and enclosed in brown oxide of iron, doubtless originated from the porphyry, and entered the joints and bedding planes of the quartzite, where they were opened by faulting. Above the shale the ore does not penetrate, the shale acting as an impervious resistance to uprising solutions. Ore bodies also occur in the Jurassic limestones below the quartzite, especially where they are penetrated by eruptive rocks.

In the eastern portion of the San Juan region some important gold deposits occur near Del Norte in the Little Annie or Bowen Mine, which appear to be a decomposed dyke of eruptive rock, containing free gold in brown iron, in the upper portion, and with depth iron pyrites also goldbearing.

CREEDE.

At the newly discovered camp of Creede, not very far from Del Norte, the fissure veins are very similar in character to those elsewhere in San Juan; they are quartz fissure veins traversing andesitic breccia and other volcanic rocks. The gangue matter in these veins is exceedingly rich in silver-bearing ore, so much so that the amethystine quartz composing the gangue or veinstone is quite in a minority to the ore, and the vein may be said to be nearly a mass of ore from wall to wall. The thick lavas of Creede rest doubtless with depth upon Carboniferous limestone or else on bare granite; the former is found outcropping at some distance from Creede, from beneath the lava overflow, and being penetrated by intrusive eruptive rocks shows signs here and there of productive ore deposits similar probably to those at Leadville. Creede is an encouraging example to a prospector, that all productive veins in Colorado have not been discovered yet, even in districts that have been pretty well tramped over. Creede had doubtless often been more or less walked over by prospectors for years before the great discovery was made, and in a year's time we may hear of several more similar discoveries in the great San Juan.

ROSITA AND SILVER CLIFF.

The next important and peculiar igneous district carrying fissure veins is that of Rosita and Silver Cliff in the Wet Mountain Valley near the edge of the prairie country in southeastern Colorado. Here a local eruption of considerable power and magnitude and of comparatively recent date has occurred. These eruptions, consisting of andesitic, rhyolitic and trachytic material have built up cones and rounded hills largely of fragmental material such as consolidated tuffs, ashes, and breccia, all of which, as at Cripple Creek, rest on granitic basement rock. From the fragmentary character of the rocks it is evident that most of the eruptions were explosive, alternating, however, with quieter flows; in some cases the dykes can be seen, where some of the lava came, at others the "necks" or throats of the volcanoes themselves filled up with volcanic boulders; of such is the celebrated Bassick Mine. The mine is in the throat of an old crater of andesite, filled with boulders of granite and andesite bedded in gravel and sand. The ore of the Bassick appears as concentric zones or shells around these boulders, as a replacement of the gravelly matrix. The entire mass has been permeated by heated waters which have decomposed the rocky fragments, depositing opaline quartz and kaolin in abundance.

The concentric shells around the boulders carry alternately several minerals,such as galena,antimony, zinc-blende, copper and iron pyrites, all more or less gold-bearing. The ore deposition in this region seems to have taken place at the close of the eruptive period, when the eruptions were

128

dying out into hot springs, fumaroles, etc., and producing great decomposition of the lava rocks. The district was not thought much of, until Mr. Bassick made his discovery in the unpromising looking throat of the old volcano, containing a formation quite anomalous, and which the regular prospector, accustomed to true, orthodox fissure veins, would have passed by as very unlikely. So it may happen to future prospectors, that some very unlikely formations may turn out great riches; hence it is well to keep a sharp lookout for everything examinable.

A STUDY OF MODERN LIVING VOLCANOES TO UNDERSTAND
THE CRIPPLE CREEK VOLCANO.

By far the most typical, instructive and important gold camp in Colorado and the West is that of Cripple Creek. To

PLATE LXXII.
Stromboli Volcano.

understand the geology of the Cripple Creek region and gold-bearing volcanic regions and rocks and their relations to the ore-deposits, a knowledge of the phenomena attending modern volcanic eruptions is necessary. Let us take that of the living volcano of Stromboli, described by Professor Judd, as throwing some light on the phenomena that may have occurred many thousands of years ago in the now extinct volcano of the Cripple Creek district.

From a point on the sides of the mountain of Stromboli, masses of vapor issue and unite to form a cloud over the mountain. This cloud is made up of globular masses, each of which is the product of a distinct outburst of the volcanic forces. At night a glow of red light appears on the cloud, increasing gradually in intensity, and as gradually fading away.

After an interval this is repeated and continues till the light of dawn causes it to be no longer visible. When we land on the island we find it built up of the "ejecta" from the volcano like a gigantic iron furnace with its heaps of cinders and masses of slag. The irregular shape and surface of the island is due to erosion removing the loose materials at some points, and leaving the hard slaggy masses standing up prominently as dykes and hard portions of lava flows, as Pisgah, Rhyolite Mt. and others at Cripple Creek do, above the eroded and more fragmentary tuffs and breccias. This great heap of cinders and slags rises 6,000 feet above the sea bottom with a base four miles in diameter; 2,000 feet above sea level is a circular depression, the crater of the active volcano.

Looking down into the crater, an outburst takes place. Before the outburst, many light curling wreaths of vapor ascend from fissures on the sides and bottom of the crater. Possibly this is the origin of some of the dyke-filled fissures of Cripple Creek. Suddenly a sound is heard like a locomotive blowing off its steam. A great volume of watery vapor is thrown up into the atmosphere, and with it a number of dark fragments are hurled 500 feet above the crater, some falling on the mountain, others back into the crater with a loud rattling noise. Those rolling down the mountain are still hot and semi-molten. This is a clue to the

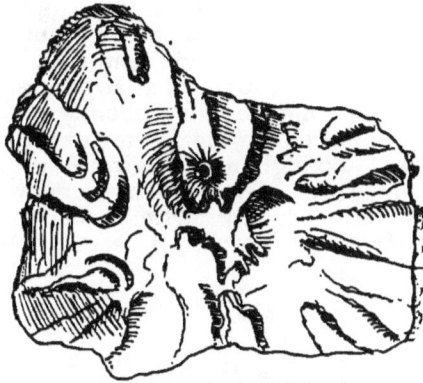

PLATE LXXIII.

Map of Island of Stromboli.

origin of the fragmentary materials composing the tuffs and breccias at Cripple Creek. The black slaggy bottom of the crater is, as we have said, traversed by many fissures emitting jets of vapor. Some of these are quite large and vary in size and number and position at different periods. From some, only steam is emitted in loud snorting puffs. In others molten material is seen welling up and flowing out—

side the crater. Such fissures when all eruption has ceased
would be found, as at Cripple Creek, sealed up with solid

PLATE LXXIV.
Stromboli Crater.

lava with a lava flow on their tops. From this liquid mass,
steam escapes in considerable quantities. Within the walls
of the fissures, a viscid semi-liquid lava heaves up and down
and churns around till at last a gigantic bubble or blister is
formed which bursts violently and a great rush of steam
takes place carrying fragments of the scum-like surface of
the liquid high into the air. At night the fissures glow with
ruddy light. The liquid matter is white hot and the scum on
it a dull red. Every time a bubble bursts a fresh glowing
surface is exposed. It is the reflection of this upon the
clouds of steam
above the mountain
that causes the fitful
glows of light we
mentioned.

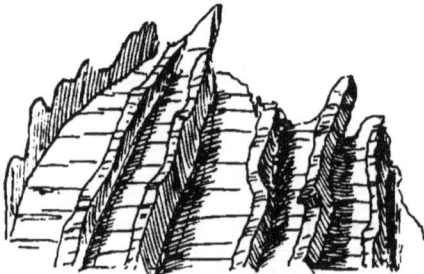

PLATE LXXV.
Dykes Cutting Beds of Scoria and Tuff in the
Wall of a Crater.

The phenomena
show there are
cracks communicat-
ing with the earth's
interior highly
heated matter be-
neath the surface,
together with great
quantities of impris-
oned water, which
escaping as steam give rise to all the active phenomena.

What is popularly supposed to be flame in an eruption is the reflection on the cloud of steam and dust, from glowing masses in the mouth of the crater. Sulphur is not, as commonly supposed, erupted from a volcano, but is formed by the union of sulphurous acid and sulphureted hydrogen issuing from volcanic vents.

A volcano is a steam vent, like a geyser, which may be called a water volcano.

ORIGIN OF FISSURES.

Some light is thrown on the possible origin of some of the Cripple Creek dykes and fissures by the eruption of Vesuvius in 1872. The bottom of the crater was entirely broken up and the sides of the mountain rent by fissures in all directions. So numerous were these fissures that liquid matter appeared to be oozing from every part of its surface and the mountain to be "sweating fire." One fissure was enormous, extending from the summit to far beyond the base of the cone. This, filled with a dyke of lava, is visible to-day. From both crater and fissures enormous volumes of steam rushed out with a prodigious roar. This roaring was from explosion of bubbles one after another, and the vapor cloud above Vesuvius, as at Stromboli, was made up of globular masses of steam ejected at successive explosions. Each explosion carried upward quantities of fragments which fell back on the mountain. All along the course of the stream of lava, volumes of steam were thrown off.

ORIGIN OF TUFFS.

The discharge of such large quantities of steam causes the atmosphere to be saturated with watery vapor, which, condensing, falls in excessive rain storms, producing mud streams formed by rain water sweeping along the loose volcanic dust and debris. In some such way, doubtless, the Cripple Creek tuffs and breccias were formed.

GASES AND MATERIALS EJECTED FROM VOLCANOES.

The most abundant of the substances ejected from volcanoes is steam, and with it many volatile materials, such as hydrochloric acid and carbonic acid, also hydrogen, nitrogen and ammonia, and at Cripple Creek fluorine gas.

These different gases at Cripple Creek had much to do with the formation of ore deposits. Volatile metals, such as

arsenic, antimony and cinnabar are erupted; these substances, issuing from volcanic vents at high temperature,
react upon one another forming new compounds, such as
sulphur. Hydrochloric acid unites with the iron in the
rocks to form yellow ferric chloride, common at Cripple
Creek, and looking like a greenish yellow sulphur. Acid

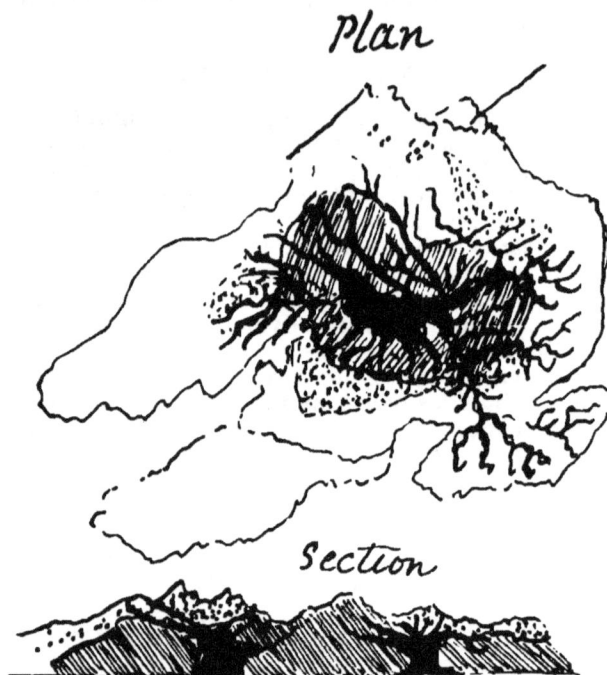

PLATE LXXVI.
Plan and Cross-Section of the Roots of a Crater.   Black=Dykes Filling Fissures.

gases change lime, alkaline and iron elements into sulphates,
chlorides, carbonates and borates, which, when removed by
rain, leave a white substance like chalk, composed of pure
silica. Beds of such material occur not far from Cripple
Creek and powdered silica in some of the mines.

The lips of fissures from which steam and gases issue are
coated with yellow and red incrustations of sulphide and
oxide of iron, such as are common in many prospect holes
at Cripple Creek.

Solid materials are ejected in vast quantities; fragments
of the rock masses through which the fissure is rent are

Microscopic Structure of Glassy Lava Showing Microlites and Crystallites.

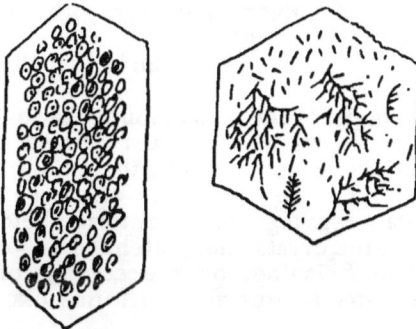

Microscopic Structure of Some Crystals Showing Microlites and Crystallites.

PLATE LXXVII.

carried upwards by the steam blast, together with other
matters far beneath the surface in a semi-fluid condition.
Hence it is that at Cripple Creek we occasionally find

fragments of red granite imbedded in the volcanic breccia torn from the throat of the volcano in its passage through the underlying granite of the region.

MINERAL AND CHEMICAL ELEMENTS OF LAVAS.

Eight chemical elements make up the mass of lavas, oxygen, silicon, aluminum, magnesium, calcium, iron, sodium and potassium. Oxygen makes up the larger proportion so that lavas are mostly oxides. Next is silicon and aluminum, giving the quartz and feldspar and silicate element.

Lavas are of two kinds, acidic and basic. Acid lavas contain eighty per cent. silica, basic forty-five per cent. The former are rich in potash and soda, the latter in lime and iron; the former are commonly light in color and weight, the latter dark and heavy. Rhyolite is an example of an acidic lava, basalt of a basic one. The andesites and phonolites of Cripple Creek are intermediate. The minerals composing these lavas are principally quartz and feldspar, together with the dark minerals, mica, augite, hornblende, olivine and magnetite.

CRYSTALS AND MICROSCOPY OF LAVA.

Many lavas are of a glassy nature, others contain many crystals, some of large size.

Microscopic sections of lavas show them to be made up of a ground mass of a glassy character, with distinct crystals set in it like plums in a pudding.

In others, the crystals are so thick that the glassy base can scarcely be seen.

Through the midst of the glass, cloudy matter is observed; a higher power shows this "nebula" to be composed of minute particles called crystallites, the embryonic forms of crystals. Sometimes we can see an attempt of these particles to aggregate into a geometrical form, sketching out the outline of the large crystal they intended to form, but were prevented from finishing, by the cooling of the glassy magma. These crystallites assume forms like ferns, hairs, spiders, etc.

In subterranean regions the conditions were particularly favorable for the development of crystals. The lavas cooled with extreme slowness, under enormous pressure, allowing plenty of time for the crystals to form.

Those lavas containing most soda and potash (acid lavas) assume a glassy condition, and these have often cooled near the surface rapidly, the more crystalline varieties slowly at great depth. Obsidian and rhyolites are glassy types, granite and some porphyries with large crystals are of the latter class, whilst andesite and phonolite may be intermediate. The latter, however, at Cripple Creek, may have cooled quickly near the surface, and the crystals are for the most part small.

Besides the natural imprisoned water, crystals in lavas are found microscopically to contain globules, sometimes filled

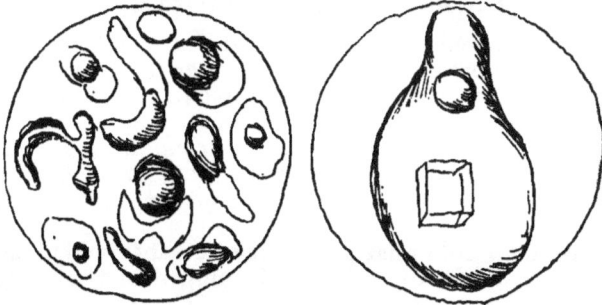

PLATE LXXVIII.

Minute Cavities Containing Liquids in the Crystals of Rock.

with gas, salt, and water, which may add to the materials for the production of steam.

ERUPTIONS OF DUST.

Steam escapes from lava so violently that the froth or scum called scoria, is broken up and scattered in all directions. This scoria like pumice is full of little holes like a sponge, due to escape of the steam in it. Such spongy scoria is found scattered over the hills of Cripple Creek. During violent eruptions a continuous upward discharge of these fragments is maintained; the cindery masses hurtling one another in the air, fall back into the vent, or are scattered over the mountain. Being often shot up again and again from the vent, they are reduced to the finest impalpable dust. They fill the atmosphere to such an extent as to bring on an " Egyptian darkness." This dust, mingling with descending rain, forms destructive mudflows, and sets or consolidates into the tufas or tuffs so abundant at Cripple Creek. When larger angular fragments are caught up and consolidated with these, the rock so formed is a breccia, as already illustrated.

Volcanic craters after having been formed, are liable to be disturbed by later eruptions. Thus the crater of Vesuvius was reduced 400 feet by a later eruption, the old crater blown up and a much vaster crater opened.

Cripple Creek also witnessed its second disturbance, after the andestic eruption had ceased, by one of phonolite lava.

FLUIDITY AND OTHER PROPERTIES OF LAVAS.

Some lavas, such as basalt, are reduced to such a state of fluidity that their streams run like water to great distances. Others are of a more viscid, mortar-like consistence, especially the acid lavas, such as those of Cripple Creek. These are apt to flow but a short distance from their source, and to build up big domes and thick masses; of such a nature seems the structure of Nipple Mountain, south of Cripple Creek.

The peculiar columnar structure often observed in basaltic lava sheets, and in a rough way developed in the phonolite of the cliff above Victor mine, is due to cooling and contraction somewhat in the same way as mud cracks are formed in a drying up pond. A block of lava isolated by these cracks assumes a polygonal form like the basaltic columns of the Giants Causeway.

During the cooling down of lava and the escape of steam and gases, deposits of sulphur, specular iron and (at Cripple Creek) fluorspar, are deposited. Specular or micaceous iron is not uncommon at Cripple Creek. Rock masses are completely disguised by these incrustations.

STRATIFICATION OF TUFFS.

Tuffs and breccias are often found stratified. The fragmentary materials in falling through the air are sorted, the finer particles being carried farther from the vent than the larger ones. Craters built up of tuffs and breccias fallen in the condition of a muddy paste, show very fine stratification.

Large cones are built up of uniformly spread layers of more or less finely divided material disposed in parallel succession. At Cripple Creek the bedding is indistinct, and often difficult to trace, the dip of stratification being still more compressed by the cross fracturing of the rocks; hence it is hard to tell whether the lines represent cross fracture cleavage, or bedding planes. In most volcanoes the stratified tuffs are cut and crossed, as at Cripple Creek, by numerous dykes running in various directions, cracks filled by lava from below.

Movements, too, have taken place subsequent to the accumulation and consolidation of the whole material as shown in Plate LXXIX, whereby the masses are faulted and fresh fissures opened in them. Faults are found in some of the mines at Cripple Creek, faulting not only the lavas, but the veins also. Cliff sections of volcanoes show alternate beds of solid lava, scoria and tuff, representing different eruptions or flows.

There seems an order and succession in the eruption of the different varieties of lava. During the earlier periods

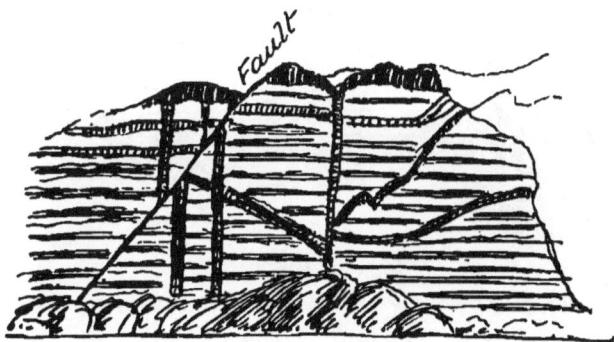

PLATE LXXIX.
Cliff Section, Composed of Alternate Beds of Lava and Scoria, Cut by Lava Dykes, and Faulted.

rhyolites, andesites and phonolites are erupted, and later basalts. This appears to be the case in the volcanic region west of Cripple Creek around Mt. MacIntyre, Thirty-Nine Mile Mt., and Black Mt. The prevalence of basalt capping the other lavas in that region, together with the greater freshness of the rocks, imply that its eruptions were somewhat later than those of Cripple Creek where basalt is not found, and where the rocks are much decomposed.

Volcanic eruptions shift their centers from time to time, making new cones along a line of fissure (for volcanoes are built upon lines of fissure). See Plate LXXX. Extinct craters are frequently filled by beautiful deep lakes. Cones rise within cones, and within great crater rings. At each successive great eruption, the old cone is blown away, and a new one formed.

Hot springs contain large quantities of silica or quartz in solution. The solution of silica is effected at the moment of its separation from combination with the alkali during the decomposition of volcanic rocks, and is favored by the presence of alkaline carbonates in the water, high temperature, and the pressure under which it exists in subterranean regions. When the water reaches the surface and is relieved from pressure and begins to cool, silica is deposited. So are the basins of geysers formed, and so the opal and hydrated quartz we find in many of the Cripple Creek veins, and in resilicated rocks.

Hot and cold springs rising in volcanic regions are charged with carbonic acid, and passing through calcareous

PLATE LXXX.
Showing Craters Found Along a Line of Fissure in the Eruption of Etna.

rocks dissolve large quantities of carbonate of lime, and re-deposit it in a crystalline form known as "travertine." Near the base of Mt. MacIntyre, west of Cripple Creek, a prospect is opened on a fissure filled with this substance.

Nearly all eruptions take place along lines of fissures (See Plate LXXX). Probably all volcanoes are located upon fissures of some kind, and even the general distribution of volcanoes over the earth's surface has been attributed to lines of fissures, as if the earth had been cracked like a glass globe. We have plenty of opportunities of seeing ancient fissures filled with lava in the numerous dykes at Cripple Creek, and in the greater volcanic region west of it; but so far no distinct volcanic craters have been found. Nevertheless it is probable that craters existed along these fissures, long since removed by erosion, or buried deep under flows and surface matter. We not unfrequently find at Cripple Creek that fissures did not all succeed in breaking through to the surface, for at some depths in the mines the apices of buried dykes are found and fissures filled by vein matter, whose outcrops do not appear at the surface. A single vein is followed from the surface and with depth two or more veins are often encountered, together with various small fissures.

Earthquakes doubtless accompanied the eruptions, and developed many smaller fissures, and further shattered the rocks. Added to this at Cripple Creek, there was the second eruption of phonolite, after the andesite had ceased. This second eruption doubtless added new fissures in the efforts of imprisoned vapors to force for themselves channels to the surface.

GASES AND SOLFATARIC ACTION.

The several stages in the decline of each volcanic outburst are marked by the appearance at the vent of certain acid gases. As the temperature at the vent declines, the nature of the volatile substances emitted undergoes a regular series of changes.

In fumaroles, sulphurous acid and hydrochloric acid abound, with sulphureted hydrogen and carbonic acid in much smaller proportions. Around these fumaroles, deposits of sulphide of arsenic, chloride of iron and of ammonia, boracic acid, and sulphur take place. Arsenical pyrites are a common associate for the ores near the surface at Cripple Creek, and many rocks are permeated with iron pyrites.

Where a volcanic vent sinks into extinction, hydrochloric and sulphurous acids are first evolved, and later sulphureted hydrogen and carbonic acid springs. Such springs are common in the volcanic districts of Colorado to-day, but we have long passed the stage of the stronger acids, which could only be expected in the pit of an active modern volcano like Kilauea. We may, however, expect to find traces left of these gases, in the rocks of Cripple Creek, such as a bleaching and decoloration of the rocks, leaching and precipitation of iron, forming those varied patterns of oxidation so common at every prospect hole; also deposits of various sulphates and chlorides, rocks deprived of iron and alkalies reduced to powdery siliceous masses.

One action of subterranean springs is the transportation of material in a state of solution and redepositing of it elsewhere, especially in lines of relief of pressure, such as fissures, shattered rocks, and decomposed rocks and zones in the rocks.

At Steamboat Springs, Nevada, metallic gold, cinnabar and other minerals have been found coating the sides of fissures from which living hot springs issue at the surface. In great volcanic foci the transfer of various sulphides, oxides and salts, which fill veins, has been effected either by solution or sublimation, or the action of powerful currents. This applies to the veins and ore deposits in question.

As the igneous activity of a district declines, the temperature of the issuing gases and waters diminishes, till at last the volcanic forces appear to have wholly abandoned the region and been transferred to another. This may have been the case with Cripple Creek and the volcanic region west of it, of apparently later date. The history of a volcanic disturbance is as follows:

First. The area is troubled by subterranean shocks and earthquakes.

• Second. The origination of fissures is indicated by the appearance on the surface of hot and carbonic acid springs and other gases.

Third. With increased subterranean activity the temperature of the springs and gases increases.

Fourth. A visible rent is formed at the surface.

Fifth. From this fissure, gas and imprisoned vapor escapes so violently as to disperse the lava in clouds of scoria or dust, or to cause it to well out in flows.

Sixth. Volcanic action concentrates at one or several points, and the ejected material accumulates from volcanic cones.

Sometimes the volcanic activity dies out entirely, leaving cones thrown up along the line of fissure. At others, some such center becomes for a long time the habitual vent for the volcanic forces of the district, and a large cone is built up.

When the height and thickness of the cone have grown great, the succeeding eruption rends the sides of the cone, producing fissures, quickly filled by lava, forming radiating dykes and surmounted by parasitic cones. The dykes of Cripple Creek may in cases represent such occurrences.

When volcanic energies can no longer raise material to the summit of the crater, nor rend the sides, they find relief by making new fissures and small cones in the country outside the main volcanic crater. The numerous phonolitic dykes in the granitic region outside of the main center at Cripple Creek may have so originated. At last volcanic energy diminishes, eruptions of lava cease, fissures are sealed up with solid lava, volcanic cones crumble away.

But still the existence of heated matter at no great depth
is indicated by outbursts of gases and vapor, formation of
geysers, mud volcanoes and hot springs. As the underlying
rocks cool down, the issuing jets of gas and vapor lose their
high temperature, diminish in quantity, geysers and mud
volcanoes become extinct, hot springs disappear, and all is
quiet.

It was in the latter or hot spring stage, that the ores were
at Cripple Creek leached from the volcanic rocks, probably
from great depths as well possibly as from the sides, and
concentrated and deposited in the fissures, shattered zones,
and decomposed rocks. The last stage is as we find things
to-day.

GENERAL SUMMARY OF PROBABLE VOLCANIC EVENTS THAT OCCURRED AT CRIPPLE CREEK.

At Cripple Creek there was a volcanic eruption in Tertiary times due probably to some mountain elevation going
on in the region of Pike's Peak or generally in the mountains.

We may assume that preluding the eruption the area was
troubled by earthquakes. Various kinds of acid and hot
springs appeared above the surface, indicating the fissuring
of the ground that followed.

At the bottom of these fractures, which may have been
numerous, molten rock appeared, giving off imprisoned
vapor from bursting blisters of lava. These shoots of steam
formed into a cloud overshadowing the area, and carried
upwards quantities of scoria and fragments, which fell back
around the orifices, forming a crater cone, or craters. These
fragments being repeatedly shot up, and falling back into
the crater were comminuted into fine dust, and fell, together
with larger angular fragments, over the surface.

The atmosphere charged with condensing steam gave rise
to heavy rain falls. The water descending the ravines,
caught up the volcanic dust and fragments, forming mudflows, the material rapidly setting into the rocks we call
tuffs and breccias.

As the first eruption at Cripple Creek was of andesite,
these are called andesitic tuffs and breccias, and constitute
the principal mineralized rock of the mining area.

These tuffs are sometimes stratified by the materials being sorted in the air by the water.

After this first eruption ceased, there may have been a

rest for a time, the lavas may have cooled and consolidated, and the region been covered by various acid and hot springs, issuing from fissures caused by the late eruption.

Then the district was a second time disturbed, this time by an eruption of phonolite, ascending through numerous rents and fissures, not only in the overlying andesite, but also in the granitic region outside of the first volcanic "focus," probably finding the old seat of action too much choked by eruptive matter.

This second eruption added many new fissures to the already shattered rocks, and gave many opportunities for the deposition of metallic and vein material deposited through the medium of gaseous and hot spring and solfataric action which followed upon the cessation of the phonolite eruption.

After the eruptions at Cripple Creek ceased the volcanic forces seem to have transferred their field of action to the area west of Cripple Creek in the Four-mile district. The rest is the history of to-day.

CRIPPLE CREEK AS A PROSPECTING FIELD.

A visitor standing on top of one of the hills like Mt. Pisgah, overlooking Cripple Creek, and glancing at the various mines and multitudinous prospect holes speckling the hills, is struck with the compactness of the mining district within the limited area of 18 square miles. In this small area all the principal mines are located, and one can ride around the entire camp in an hour or two. Outside of this area, there are as yet no mines of importance, though prospect holes may be found for a circuit of many miles.

ANDESITIC AND GRANITE AREAS.

He will observe that the principal mines are located on the round smooth hills, on their tops, slopes and on the gulches, where the vegetation is mostly grass and quaking aspen. These too are within a sort of natural rampart of more rugged hills wooded with pine. In these outlying hills, only a few scattered prospects are visible. The reason for this is to be found in the geology of the region, and the differences between the areas occupied by andesitic breccia and granite. The rounded grassy aspen-covered hills representing the andesitic breccia carry most of the ore bodies, and the principal mines are restricted to them. The more rugged hills, covered with fir trees, represent the granite area, and in them for the most part are few mines of importance, though many likely prospects are opened upon dykes of phonolite, which, so far as known, does not as a rule seem to be so productive a rock as the andesite.

There are intermediate areas, such as that of Battle Mt., characterized by the presence of both andesitic breccia, phonolite dykes, and granite, in which are some of the richest mines of the district, such as the Independence, Portland, Annie Lee and others.

It will appear how important and useful a geological survey is of such a region, a fact not always recognized by practical miners. If the ore bodies are mainly associated with the particular rock called andesitic breccia, it is well for them to be able to recognize that rock, and ascertain the limits of its area.

SIGNS THAT LEAD TO PROSPECTING.

The next thing that strikes the observer, is the prodigious amount of prospecting holes and prospecting trenches, the latter being particularly common. He may ask, what was there in the general appearance and character of this district that led the "eagle eyed" prospector to suspect the existence of ore bodies in it, or that it was "a kind'er likely looking place"? Again, how is it that it was so long overlooked by the "eagle eyed," especially when so easily accessible?

On general principles, in past years, miners in Colorado, after the Leadville and Aspen excitement, were more on the lookout for silver than gold; they looked therefore for rocks like those of Leadville, with contacts between porphyry and limestone, and every limestone ledge in the country was ransacked. Silver was rarely found in volcanic lava rocks, except perhaps in the great San Juan region, and miners thought as little about prospecting unpromising looking hills of lava, as they would the basaltic caps of the table mountains on the plains. Again, gold leads do not show their ore on the surface like some silver-lead veins. There is nothing perhaps but a little seam of rust that might occur almost anywhere, and in any kind of rock. Hence lava districts of somewhat recent origin, were overlooked, rather than looked over. The discovery of the gold-bearing properties of the Cripple Creek lavas, together with the increased thirst for gold, turned the tables, and

now throughout Colorado, every lava formation is being
prospected with as much zeal and indiscriminateness, as
were the limestones in the Leadville days. The prospector
now needs to know volcanic lavas at sight, to distinguish
varieties, and to know all he possibly can about their
origin, varieties and mode of occurrence. Hence the importance we gave to the subject in the preceeding remarks
on volcanoes. A prospector *now* would at a glance consider the area about Cripple Creek as worth looking over;
and the geologist would consider it a very likely place, not
merely from the presence of the lavas, but mainly from the
great decomposition of the rocks, and the evidence of the
presence of past solfataric action.

DIFFICULTIES IN PROSPECTING.

But the "eagle eyed" one did not entirely overlook this
district in the past, for some years ago he was sufficiently
prepossessed with the appearance of things to drive a
couple of short tunnels in Arequa gulch, and narrowly escaped becoming a millionaire. What troubled the prospector was, that though he found the hills covered with an
extraordinary amount of "float," he could not trace this
float to any ledge or rocks "in place." For the most part
the hills were grassed over, or covered with vegetation; and
through the turf were very few outcroppings of a likely
kind, so far as he could see. There were no prominent
quartz veins, or zones deeply impregnated with iron, hence
he gave up the region, mentally wondering where on earth
all this rich float could have come from, perhaps solacing
his mind by one of his igneous, brimstony theories that it
had been scattered over the country from a distant volcano,
or washed there by flood or glaciers from some unknown
distant region. The former theory after all was not far
from the truth but the absence of all rounding and smooth-
ing of the fragments of float precludes the latter hypothesis. Evidences of former glaciation are remarkably
absent from the vicinity.

THE REGION IMPREGNATED WITH ORE.

To those who have studied Cripple Creek of to-day, the
source of this "float" is no mystery. Little, if any of it,
has been broken off from orthodox quartz fissure veins, or
even extracted from well defined ore zones. The fact is,

that the whole andesitic area is more or less impregnated
with the precious metals, and the float on the surface is
little more than the surface debris of the general underlying rock. There is scarcely a stone that you may kick with
your foot over the entire area, but what will show some
trace of gold, On one hill an experienced mining superintendent told me, that for an experiment, he went around
with a wagon and picked up the "float" almost at haphazard, and it averaged 22 dollars in gold. That such a
"floaty" region should receive attention some day is not to
be wondered at, and we believe Colorado Springs men were
amongst the first to give it serious attention by opening
holes and prospecting trenches almost at random, resulting in important discoveries. As a rule even after this, the
best mines were discovered by mere chance and guess
work, or by plodding but blind prospecting, something
like the Leadville prospector who in early days had all
Leadville before him to prospect, but did not know where
to begin, till sitting down under a tree eating his lunch, he
saw a squirrel scratching in the ground ; he accepted the
happy omen and "went down," so the story goes, and of
course "struck it rich ; " so we understand the Pharmacist
and many other now noted mines were discovered at Cripple
Creek.

MODE OF PROSPECTING.

This absence of surface outcrops or visible leads, when
the "rush" came, led to indiscriminate and abundant
prospecting which has been kept up till the present time,
hence, the extraordinary freckling of the hills with prospect
holes and trenches.

Sometimes they would select any piece of land they
thought, for some reason or other or without any reason
at all, likely, and go to work to punch holes and dig
trenches all over it to find something. In this way they
frequently came across enough signs to warrant putting
down a prospect hole, and holding the claim and then went
on " to pastures new."

CHARACTER OF FLOAT AND OTHER SURFACE SIGNS.

As we have said, the whole region is covered with float.
This float is usually a somewhat porous piece of lava, or
andesitic breccia, or tuff, stained with yellow, brown, or red
oxide of iron, sometimes in patterns or concentric rings. It

is often found to be honeycombed when broken with a hammer. There is no visible ore, but an assay will most likely show traces of more or less gold. Again, a species of

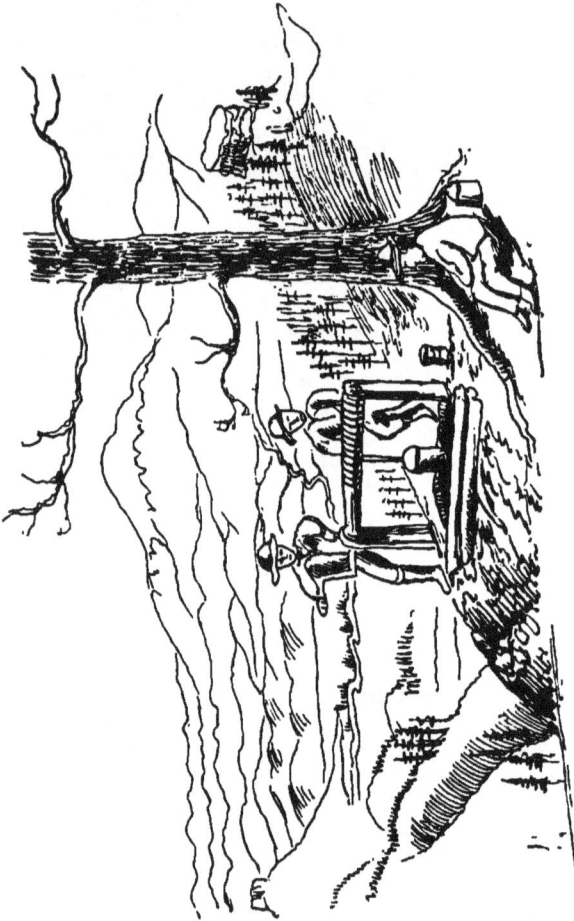

PLATE LXXXI.

Prospectors Opening a Prospect in a New Region West of Cripple Creek.

red porphyritic granite has been desilicated and robbed of many of its crystal constituents, and left as a porous skeleton of a rock by the action of gases and springs. The pores in this are often occupied by oxide of iron, or even by crystals of fluorspar. This is a likely kind of float. Honeycombed rusty rock with quartz crystals is a likely float, both of these representing the action of mineral hot springs. At rare intervals we may see a little of this oxidized rusty rock in place protruding from under the grass, and if so, there is sure to be a prospect hole alongside of it.

Bold outcrops of lava rock are comparatively scarce and when they do appear, as in the cliff above Victor mine, Mt. Pisgah, Bahr, and Rhyolite peaks, the rock is apt to be so hard as to preclude the probability of much ore deposits in it.

Pieces of rock or float stained a violet purple color by fluorine are considered a good sign of an ore body not far off, this fluorspar being found characteristic of some of the richest veins in the camp ; and fluorine gas was doubtless connected with the deposits of ore matter, especially of the tellurium, the present matrix of the gold in the deeper parts of the mines.

Pyrites is not usually found on the surface till the rock is broken open, and tellurium in little silver scales and spots, not till considerable depth is attained. But free gold may be found in surface float, and from the grass roots down, and in the early development of a mine, in the oxidized upper portions, associated with iron oxide and black manganese or " psilomelane."

Micaceous or specular iron, is seen in some prospect holes ; and localities marked by evidences of past hot spring action, such as the appearance of botryoidal chalcedony or opal should be prospected. A common and curious marking in some of the bleached volcanic lavas is that of an imitation of trees, ferns and mosses, popularly called " photographic rock," scientifically " dendrite " or tree rock.

This remarkable imitation of nature is due to crystallization of solutions of manganese, and may be compared to fern-like appearances on a frosty window-pane in winter, which are certainly not of organic origin, or in anyway connected with the processes of photography. These dendritic markings may or may not be considered as signs of ore. Similar markings are very common in the porphyries of Leadville overlying the silver deposits.

SURFACE PROSPECTING OF A MINE.

In some of the surface discoveries of mines, when a considerable area, covered by a blow-out of iron oxide as-

148

sociated or not with purple fluorspar, has been found to run well in free gold, the ground is prospected and developed to the depth of a few feet, and over a certain area, with plows and scrapers, the material so obtained being sent wholesale to the stamp-mill and often giving rich returns. The object of this work is not merely to get all the values out of this rich float, but in hopes of uncovering the vein or veins of which it is the oxidized cap or blossom. This was the way in which the Deerhorn mine was opened up, and its veins discovered on Summit Hill, The ground on the top of the hill is observed to have been "gophered" in all directions like the catacombs to a depth of about 20 feet, and over an area of a square acre or so. This was done partly to gather up and collect the rich float which was found scattered over the hill and partly to discover the leads in place.

This rich float was stained with purple fluorine, and upwards of 25,000 dollars' worth of gold was obtained from this, the material being dug up by plows and scrapers, before the subsequently discovered veins were found or worked.

In the case of the Anaconda mine on Gold Hill, the outcrop of a dyke of andesite was discovered on the hillside covered with an oxidzed crust carrying gold. The owners developed this by an open quarry, about a hundred feet in length and 40 to 50 feet deep, from which they extracted the bonanza which made this mine at its outset so celebrated, and later proceeded to uncover the dyke on the surface, to a depth of about 20 feet along the entire length of their claims, but nothing comparable with the bonanzas of the first quarry has been found since in extension or depth.

RICHNESS WITH DEPTH, ETC.

Many of the mines shipped their best ore from the grass roots and upper oxidized portions of the veins, which contained free gold and were free milling. With depth some of these mines have not done nearly as well, especially when they reached the unoxidized zone, away from surface influences, and the ore was found wrapped up in tellurium or iron pyrites.

The palmiest days of many a gold camp are its earliest days.

SUGGESTIONS TO PROSPECTORS.

In the more productive area the prospector will do well to keep to the andesitic breccia, and follow the signs we have mentioned. Outside of this area his course may be a little different, as then he is in the granite district, and looks out for the appearance of dykes of phonolite, rarely more than a few feet, though sometimes many yards, in width, and easily distinguished from the red granite by their light gray or white color. These dykes do not often appear out-cropping in the granite cliffs, but are more commonly to be found buried beneath the debris and grass of the slopes. On these he may find no indication and trust to haphazard trenching; or a few stray pieces may lead him to the spot.

The more rusty, oxidized and decomposed the phonolite, the more likely it is to carry gold; at times he may find ore

PLATE LXXXII.

ection Moose Mine Vein, Raven Hill. 1. Country Rock Breccia. 2. Yellow Jasper, with Cavities of Quartz Crystals. 3. Blue Grey Jasper, with Seams of Quartz and Iron containing Gold.

and free gold in the dyke itself, but more often at its con-tact, on one or both sides, with the granite. There he is likely to find a crevice filled with clay or iron-oxide, carry-ing seams and cavities lined with quartz crystals or stains of purple fluorspar,

Sometimes he may find the coarse granite, as in the case of the Independence mine on Battle Mt., just at the contact with the dyke of lava, to be very rotten, much honeycombed and robbed of many constituent minerals, and these, by re-placement with metal, may yield him the richest ore. Again, the dyke between walls may be reduced to a blue or yellow jaspery clay, with a vertical lamination or cleavage, the lines of cleavage filled with quartz and iron oxide (See Plate LXXXII); in such lines he is apt to find the richest ore.

After opening a prospect, the ore signs, consisting of stains of oxide of iron and manganese, instead of pursuing an even or regular course are apt to scatter amongst the infinite number of crevices shattering the rocks, no one little lead being of sufficient richness to follow with profit, and the whole body between walls scarcely paying to work. The ore signs often follow a very uneven course, now lying upon a fairly defined wall, then running for a distance into one wall or other, or again following the main course of the creviced lava breccia between walls, now in pockets and crevices, again scattered, or again impregnating the porous and decomposed rock. There are very few true, well-defined veins in the camp; the ore rather impregnates certain ill-defined, shattered zones of rocks between certain ill-defined boundaries called walls. At others the ore occupies narrow cleavage planes in the rock, of which there may be two or three in a mine, some of them productive, others very little so. Ore bodies in the harder or more compact rocks, such as the Buena Vista and Victor mines, are apt to have something more like defined veins and defined walls. In some cases surface signs have been poor, and with depth have done well; the exact opposite has often been the case. Some mines have been good from bottom to top, but we have to be careful here, as in most gold camps, of the old fallacy of "richness with depth." There is little more criterion for this than in other camps, and many a once famous mine is looking vainly with depth for its lost bonanza, though in other respects doing fairly well.

As regards the granite itself, we have heard of few ordinary quartz fissure veins unaccompanied by lava intrusions proving productive.

The fine grained, red, eruptive granite on Barnard Creek, north of Cripple Creek, has shown a promising ore body in a lava dyke in the granite, which, singularly enough, produces a fine grained galena, rich in gold. Galena is quite a rare ore in Cripple Creek. Green carbonate of copper stains appear at times in the schists and gneisses, but none so far productive.

The railroad from Canyon City to Cripple Creek did some good prospecting work in the granite area, its cuttings exposing quite a number of phonolite and other dykes, together with some granitic veins.

Outside of Cripple Creek, in the great volcanic area to the north, between Cripple Creek and South Park, is a fair prospecting field. The rocks are mainly granites, rhyolites, trachytes, andesites and basalt, the products, as at Cripple Creek, of a series of volcanic eruptions, of which the latest appears to have been basalt, which commonly caps the other and lighter colored lavas.

The rocks in this region are for the most part less decomposed than those at Cripple Creek, which is not so favorable a sign. Here the prospector should look out for all signs of decomposition, such as we observed at Freshwater district, a not unlikely spot. The very hard, massive rocks are not likely to be productive, such as the hard black basalts. The lighter colored and more decomposable lavas offer a better chance.

Centers of eruption, such as relics of old craters and dykes from which these different lavas issued, should be sought for and prospected. Balfour, a small mining camp at the north of this area, is established among granite and eruptive rocks, which have been found to be mineralized by pyrites. The granites here have several fissure veins and dykes in them, showing considerable disturbance to have taken place in that neighborhood. The low hills in which the prospect holes are located are capped with basalt, apparently resting on volcanic tuffs and other lavas. So far, nothing very productive has been found, though here, as elsewhere, much is hoped for with depth. Singularly enough, in one of these veins in lava, we noticed a tarry substance or inspissated bitumen in the cavities of the rock, an unusual occurrence in fissure veins or in volcanic rock.

CHAPTER XII.

ORE DEPOSITS IN SEDIMENTARY ROCKS.

BLANKET ORE DEPOSITS, CONTACT DEPOSITS.

This great second class of ore deposits, occurring principally in Paleozoic limestones at contact more or less with intrusive sheets of porphyry, is mainly represented in Colorado by the Leadville and South Park mining district, the Kokomo and Red Cliff districts, and the Aspen and Gunnison districts, though locally here and there, wherever Paleozoic strata accompanied by igneous rock may be exposed, silver mines may be found. We will begin with Leadville and South Park as primarily instructive and typical.

SOUTH PARK ORE DEPOSITS.

The basin plain of South Park is underlaid by sedimentary rocks from the Cambrian below, to the Upper Cretaceous on top. These strata slope up to the crest of the Mosquito range on the west, where they become violently folded and faulted and eroded.

The mineral developments are on the slopes of this range on both sides of it.

The order of succession of strata forming the structure

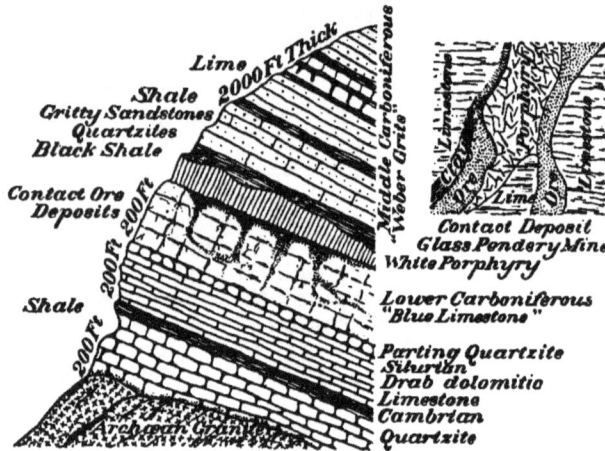

Lime
2000 Ft. Thick
Shale
Gritty Sandstones
Quartzites
Black Shale

Middle Carboniferous
Weber Grits

Contact Ore Deposits
200 Ft.

Shale

200 Ft.

Archæan Granite

Lime

Contact Deposit
Glass Pondery Mine
White Porphyry

Lower Carboniferous
"Blue Limestone"

Parting Quartzite
Silurian
Drab dolomitic Limestone
Cambrian
Quartzite

PLATE LXXXIII.

Section of Leadville Cliff.

and cliffs of the range and resting on the granite, is as follows, beginning with the lowest:

Feet thick.

Cambrian quartzite...200
Silurian drab limestone (dolomite)..............................200
Lower Carboniferous blue limestone.............................200
Middle Carboniferous sandstones and quartzite (Weber grits)..2,000
Upper Carboniferous limestones, reddish sandstones 1,000

Total...3,600 to 4,000

These formations have been traversed by eruptive quartz-porphyry and porphyrite dykes and intrusive sheets. The dykes occur principally in the Archæan, but the intrusive sheets are many and are spread out between the quartzites and limestones of the Cambrian, Silurian and Carboniferous.

The connection between the eruptive masses and deposition of ore is very marked. The ore bodies are a concentration of the metallic minerals originally disseminated through the mass of these eruptive porphyries and deposited along their plane of contact with the sedimentary beds, and by metasomatic substitution extending more or less into the mass of the latter.

On mountains Lincoln and Bross, in the principal mines, the ores are mainly argentiferous, yielding galena and its products of decomposition, viz., carbonate of lead (cerussite) and sulphate of lead (anglesite) with chloride of silver. Barite (heavy spar) is a common gangue or veinstone especially in the richest parts of the mine. Iron pyrites decomposed and passing into a hydrated oxide of iron, together with a black oxide of manganese, give to the ore its rusty and black color.

The deposits occur in irregular bodies or pockets often of great size, in the blue limestone, near its upper surface, but not always easy to find or follow. This limestone was originally covered by a sheet of quartz-porphyry which has been locally removed from the ore deposits, but exists in the peak. This porphyry, generally recognized by its large feldspar crystals is called Mt. Lincoln porphyry and is quite common and characteristic of Western Colorado. In the Dolly Varden mine the ore occurs in the limestone at contact *with a vertical dyke of white quartz-porphyry.*

In the Fanny Barrett mine, on Loveland Hill, rich deposits of galena and anglesite occur in a vertical fissure (probably a gash vein) crossing the hill from side to side and traversing the Paleozoic strata at right angles to their dip, but probably not entering into the underlying granite. This mine was discovered by noticing little pieces of iron following a general line across the hill.

In Buckskin Gulch the Phillips mine is an immense mass of gold-bearing iron pyrites, deposited, in beds of Cambrian quartzite *near a dyke of quartz-porphyry.* This mine was discovered by its rusty outcrop being exposed along the edge of the stream. At first this crust of iron oxide was loose enough to be panned for gold with good success by the old timers, and afterward milled. But when the hard pyrite set in, the ore was found to be too low grade to pay for roasting and smelting, and for many years lay idle. The Criterion in the cliff above this consists of *large caves* in Cambrian quartzite, still partly occupied by oxidized gold-bearing iron ore, and galena-bearing silver *close to a porphyrite dyke.*

The London mine in Mosquito gulch is peculiar and instructive as being involved in the great London fault. There are two strong veins or deposits of pyrites carrying both gold and silver, the gangue of one is quartz, the other

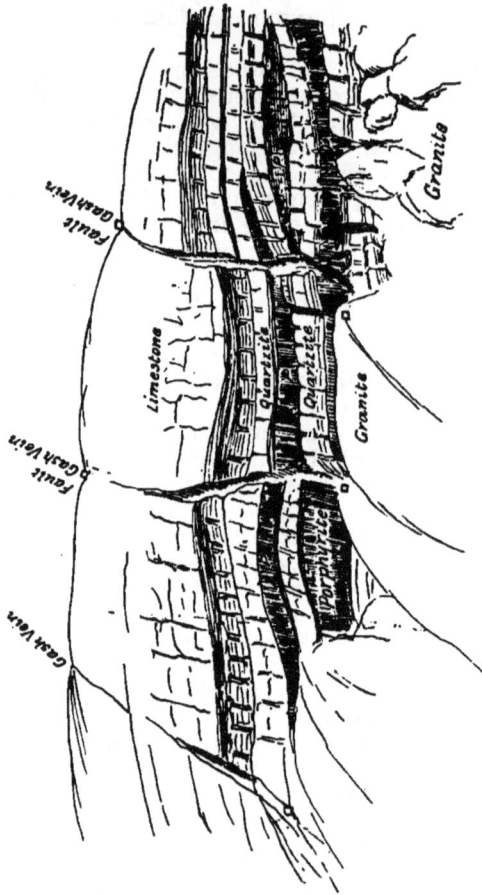

PLATE LXXXIV.

Silver Lead Gash Veins in Faults in Paleozoic and Porphyrite Rocks, Buckskin Canyon, South Park. These Veins do not extend down to the Underlying Granite.

calcite. They occur in the limestone in connection with an *intrusive bed of white porphyry*. These deposits stand in a vertical position, the beds containing them having been turned up abruptly against the great London fault, by whose movement the Archæan granite rocks forming the eastern half of London Mt. are brought up into juxtaposition with the Silurian and Carboniferous beds at its western point.

Going south along the Mosquito range the intrusive *porphyries diminish in extent and with them also the mineral deposits.*

The Sacramento mine is a good example of a "pocket" mine. Rich bodies of galena and rich decomposed ores have been found at uncertain intervals in a series of pockets or cavities. Some of these pockets or cavities are empty, and lined with modern stalactites, others contain loose

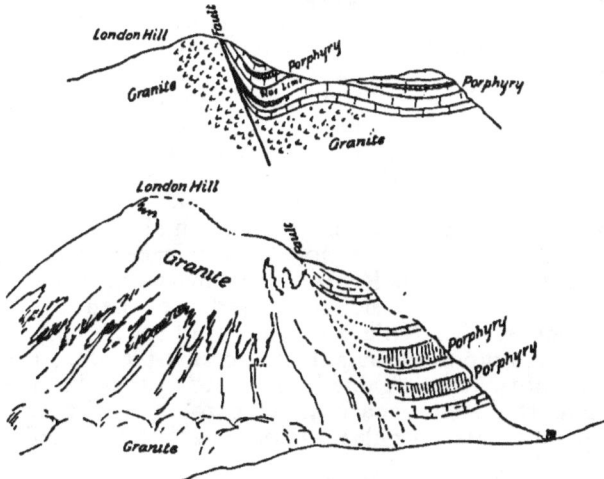

PLATE LXXXV.

The London Mine Fault.

sand, with pebbles of rich ore, others are quite full of rich ore deposits. These deposits are difficult to follow with any degree of certainty, and much of the profits made in the rich pockets has been used up in blindly "gophering" after other pockets. From some of these chambers open fissures or joint planes ascend to the surface. The limestone was originally capped by a porphyry which has since been eroded off. This porphyry doubtless supplied the ore.

LEADVILLE DISTRICT.

The western boundary of this district is the Sawatch range of Archæan granite. The slope of the Mosquito range in the east and the hills on the north, forming the water shed between the Grand and Arkansas Rivers, have a basis of Archæan granite and gneiss more or less covered by patches and remnants of the Paleozoic formations, *i. e.*, Cambrian, Silurian and Carboniferous, which have escaped erosion.

Their lower position relative to corresponding beds on the eastern or South Park side of the Mosquito range is due in part to faulting, and in part to folding of the beds.

Within these Paleozoic formations, these beds of quartzite and limestone, *there is an enormous development of eruptive rocks*, principally quartz-porphyries partially occurring as dykes but generally as immense intrusive sheets following the bedding plane of the sedimentary rocks.

Glaciers have been at work also in this neighborhood. A huge "mer de glace" occupied the great valley of the Arkansas to whose bulk numerous side glaciers contributed; these glaciers have carved and sculptured the mountains. In the flood period following the first glacial epoch a lake was formed occupying the head of the Arkansas Valley. The stratified gravel and sand beds which were deposited at the bottom of this lake now form terraces bordering the valley of the Arkansas River. These beds, known as "wash" or placer grounds, yield gold and are open to further development. Leadville is the center of the mining district, the ores are argentiferous galena and zinc-blende, They are smelting ores. Their value is increased by their having been oxidized, the lead occurring as carbonate, the silver as chloride in a clayey, or else silicious, mass of hydrated oxides of iron and manganese.

The ore is principally confined to the horizon of the "blue" or Lower Carboniferous limestone, covered by an intrusive sheet of "white Leadville quartz-porphyry." The ore bodies occur not only at the immediate contact of these rocks, but extend down in irregular pockets and chambers into the mass of the limestone, sometimes to a depth of 100 feet. Sometimes the ore completely replaces the limestone between two sheets of porphyry, as in the "Col. Sellers mine," Chrysolite, Little Pittsburg, and on Fryer Hill. A few ore bodies occur, carrying more gold than silver, found at other horizons, usually as "gash" veins running across the stratification or along bedding planes. Such are the Colorado Prince in quartzite, the Tiger and Ontario in the Weber grits of the Middle Carboniferous.

The "Printer Boy," one of the oldest mines, has produced a good deal of gold, found as free gold associated with carbonate of lead and galena, passing down, as is usual in gold mines, into unaltered auriferous iron and copper pyrites, which occur in a body of quartz-porphyry along a vertical cross-joint or fault plane in the porphyry. The gangue is a white clay resulting from decomposition of the quartz-porphyry and though the clay ore is rich, it shows no minerals to the eye.

The Paleozoic formations, together with the intrusive porphyry sheets sandwiched in between them, have been compressed into gentle folds, and where the fold was at its greatest tension, a series of parallel faults have occurred having a general north and south direction; their uplifted side is generally to the east.

The prevailing eruptive rock is the "white Leadville porphyry," occurring generally above the blue limestone but also in places below it and at other horizons.

There are also other intrusive sheets of different varieties of quartz-porphyry. The ground is generally buried beneath a hundred feet of glacial moraine material, locally called "wash."

The general geology of the South Park and Leadville region has been so elaborately traced by the labors of the U. S. Geological Survey that we cannot do better than give an abstract of their report in this connection:

MOSQUITO RANGE.

A study of this range is necessary to the understanding of the Leadville ore deposits, which occur on its western side. It comprises a length of 19 miles along the crest of the range, and in width including its foothills bordering the Arkansas Valley on the west, and South Park on the east, a slope, in one case of 7½ miles, and in the other of about 9 miles. All of it is about 10,000 feet above the sea level.

The range has a sharp single crest trending north and south. To the west this crest presents abrupt cliffs descending precipitously into great glacial amphitheatres at the head of the streams flowing from the range. Mts. Bross, Cameron and Lincoln constitute an independent uplift. The abrupt slope west of the crest is due to a great fault extending along its foot, by which the western continuation of the sedimentary beds, which slope up the eastern spurs and cap the crest, are found at a very much lower elevation on the western spurs. The jagged step-like outline of the western spurs is due to a series of minor parallel faults and folds.

The secondary uplift of Sheep Mountain on the eastern slope is due to a second great fold and fault.

The elevation of Mount Lincoln is the result of the combination of forces which have uplifted the Mosquito range and those which built up the transverse ridge separating the Middle from the South Park.

The range has been sculptured by glaciers into canyons, and the Arkansas valley is covered with horizontal terraces representing the distribution of material by waters, on the melting of the glaciers.

In the seas of the Paleozoic and Mesozoic eras which surrounded the Sawatch islands, some 10,000 to 12,000 feet of sandstones, conglomerates, dolomitic limestones and shales were deposited. Towards the close of the Cretaceous, eruptions occurred by which enormous masses of eruptive rock were intruded through the Archæan floor into the overlying sedimentary beds, crossing some of the beds, and then spreading out in immense intrusive sheets along the planes of division between the different strata.

The intrusive force must have been very great, since comparatively thin sheets of molten rock were forced continuously for distances of many miles between the sedimentary beds.

That the eruptions were intermittent and continued for a long time is shown by the great variety of eruptive rocks found. That this eruptive activity preceded the great movement at the close of the Cretaceous, which uplifted the Mosquito range as well as the other Rocky Mountain ranges, is proved by the folding and faulting of the porphyry eruptions themselves.

In the period intervening between the close of the Cretaceous and the deposition of the Tertiary strata, during which the waters of the ocean gradually receded from the Rocky Mountain region, the pent-up forces of contraction in the earth's crust, which had been long accumulating, found expression in dynamic movements of the rocky strata, pushing together from the east and the west the more recent stratified rocks against the relatively rigid masses of the Archæan land, and thus folding and crumpling the beds in the vicinity of the shore lines.

The crystalline and already contorted beds of the Archæan doubtless received fresh crumples in this movement.

A minor force also acted north and south, producing gentle lateral folds along the foothills at right angles to the trend of the range. These movements were not paroxysmal

or sudden and violent, but protracted for an enormous lapse of time, and appear to be continued in diminished force up to the present day.

MINERAL DEPOSITION.

It was during the period intervening between the intrusion of the eruptive rocks and the dynamic movements which uplifted the Mosquito range, that the original depositions of metallic minerals occurred in the Leadville region in the form of metallic sulphides, though now they are found largely oxidized and in other combinations. They were derived from the eruptive rocks themselves and are therefore of later formation than they. Their having been folded and faulted with them shows that they must have been formed before the great Cretaceous uplift, and therefore they are older than the Mosquito range itself. The deposits were formed by the action of percolating waters taking up certain ore materials in their passage through neighboring rocks, and depositing them in more concentrated form in their present position. This may have taken place while the sedimentary beds were still covered by the waters of the ocean, and the waters therefore may have been derived from it, or the area of the Mosquito range may have already emerged from the ocean and the waters have been estuarine. The uplift of the Mosquito range consisted of a series of folds fractured by faults. The crest is formed by the Mosquito fault, another parallel fracture is the London fault. The greatest movement is towards the center or Leadville region, dying out at either end north and south : the greatest displacement is 10,000 feet. Whatever cliffs may have originally been formed by this faulting have been planed down by glacial erosion.

ORIGIN OF LEADVILLE ORE DEPOSITS.

The ores are deposited for the most part in the blue limestone of the Lower Carboniferous. As the ores were deposited by water solutions, the soluble limestone beds would be more easily acted upon by solutions than the sandstones and shales composing the other rocks of the neighborhood, which are less susceptible to percolating water. The Paleozoic formations in America are the principal repositories for lead and silver ores, not by reason of their geological age, so much as by their containing such a quantity of soluble limestones and being physically as well as chemically favorable for the reception of mineral solutions.

The physical structural conditions of Leadville are particularly favorable to the concentration of percolating waters in the blue limestone. Great intrusive sheets of porphyry follow the limestone persistently, principally on its upper surface. This porphyry is very porous, and full of cracks and joints, affording ready channels for water from above, and also channels for ascending water from below, along the walls of the fissures, through which it is erupted. Such waters passing through a medium of different composition would be ready for a chemical interchange with the limestone.

COMPOSITION OF ORES.

The ores were deposited originally as sulphides. This is shown by the fact that the oxidized ores near the surface pass down with depth into sulphides. In Ten-Mile district these oxidized ores are seen to result from the alteration of a mixture of galena, pyrite, and zinc-blende. There is very little gold in the average Leadville ores; what little there is comes from the Florence mine (native gold), and from others where it is associated with pyrites. It is usually associated with porphyry rocks, and a porphyry commonly called pyritiferous porphyry shows gold to exist diffused through the pyrites disseminated through its mass.

Silver occurs as chloride, a secondary condition, its original condition probably being sulphide.

Lead occurs as carbonate and sulphate and, deep in the mines, as sulphide. Specimens are common of galena nodules surrounded by a thin coat of sulphate, and that again by a coat of carbonate, showing the order of transition from sulphide to sulphate and thence to carbonate.

In the iron mine native sulphur occurs as an alteration product of galena.

Iron and *manganese* constitute rather a gangue material than an ore. They are hydrated oxides and protoxides. The iron was originally deposited as sulphide or pyrites, but has been wholly transformed by oxidation.

Zinc is not common, but occurs as calamine (zinc silicate) in needle-like hairs and white crystals in cavities in the mines. Its original form was zinc-blende (zinc sulphide), as shown in the Ten-Mile district.

The earthy minerals, alumina, lime, silica and magnesia, are in fair proportions, as might be expected from ores which are a replacement of limestone in close connection with porphyry. The alkaline element among the ores might also be traced to the influence of the latter rock.

The agents of alteration were surface waters, which contain everywhere carbonic acid, oxygen, organic matter, chloride of sodium (common salt), and phosphoric acid. The rocks through which these waters passed, such as porphyries and limestones, were found to contain phosphoric acid and chlorine, while organic matter exists in the blue limestones ; and in the overlying shales and sandstones are many carbonaceous beds and even beds of coal. Water passing through these rocks would take up all these elements and be ready for chemical reactions.

Galena (lead sulphide) is much richer in silver than its alteration product, carbonate of lead, or cerussite. On Carbonate Hill the carbonate averages 40 oz. silver, the galena is 145 oz. to the ton. But galena is harder of treatment.

Silver is found at times disseminated through vein matter and country rock, without the presence of lead, proving that during alteration silver was removed farther from its original condition and more widely disseminated than lead.

Outcrop deposits have proved in many cases richer than those at depth. The deposits near the surface have been the refined, concentrated remains of larger bodies gradually removed by erosion, as the alteration by surface waters went on. The baser and more soluble metals have thus been removed in solutions, leaving behind the more valuable and perhaps less soluble metals in new and richer secondary combinations.

"Kaolin" or "Chinese talc," which occurs both along the line of contact and between the porphyry and limestone and also in the heart of the ore deposit, is a decomposition product from porphyry. It consists principally of hydrated silicate of alumina derived from the feldspars of the porphyries, perhaps at the time when acted upon by sulphurous waters, which brought in the original ore deposits.

Calcite occurs incrusting recent crevices and lining recent cavities.

Barite is common, generally associated with chloride of silver and manganese and is *locally recognized as a sign of rich ore.*

MODE OF FORMATION OF LEADVILLE ORE DEPOSITS.

The ores were deposited from water solutions by a metasomatic interchange, *i. e.,* substance exchanged for substance with the limestone ; and lastly or originally as sulphides.

Mineral matter is carried from one place to another within the earth's crust by heat and water, or these combined. Metasomatic interchange of metal for limestone and the removal of dolomite could only have been produced by water. The ores were *not* deposited in *pre-existing cavities*, but are a replacement of the country rock, *i. e.*, dolomitic limestone.

The ores grade off gradually into the material of the limestone, with a definite limit, as would not have been the case if the limestone had been previously caverned. The only limiting outline to the ore bodies is that formed by the contact porphyry.

Fragments of unaltered limestone are found entirely enclosed within the ore bodies, and ore bodies often occupy the entire space for long distances between two horizontal sheets of porphyry, which space further on is occupied by the limestone. This is well seen in Colonel Sellers mine. Examination of ores and veinstone shows lime and magnesia not in the crystalline condition they would have, had they been brought into a pre-existing cavity and deposited, but in the same granular condition in which they exist in the country rock.

The deposits in rocks other than limestone consist of metallic minerals and of altered portions of the country rock, in which the structure of the latter can sometimes be still traced, and are not the regular layers of matter foreign to the country rock, which results from the filling of a pre-existing fissure or cavity by materials brought in from a distance and deposited along the walls.

In the Ten-Mile district the arrangement of the particles of the original rock is frequently seen to be preserved in the metallic minerals, which maintain a certain parallelism with the original bedding planes in the lines defined by minute changes in these minerals.

The common characteristic of caves which have been dissolved out of limestone is, that their walls are coated with a layer of clay which has been left undissolved by the percolating waters, and these walls have a peculiar surface of little cup-shaped irregularities from which also stalactites frequently hang. There is also an accumulation at the bottom of the cave of fragments of limestone, fallen from the sides of the roof. None of these characteristics are found associated with the ore replacements.

Also, when mineral matter is deposited in "pre-existing cavities" it takes the form of regular layers parallel with the walls of the cavity, as is beautifully shown in geodes lined with a succession of zeolites or with layers of chalcedony, opal and quartz.

No such successive arrangement in layers is found in the Leadville ore bodies.

Again, could such large, open cavities have existed for long distances without support between the layers of porphyry? Why did not these porphyry sheets close together? And further, how could such extensive cavities have been formed and kept open under a pressure of 10,000 feet of rock, which the geology of the region shows to have existed above the deposits at the time they were being formed? Such cavities as we do find in the region are all of very recent origin, cutting through both limestone and ore bodies, and have been hollowed out by surface waters more recent even than those which produced the secondary alterations in the ore bodies.

The ore deposits of Ten-Mile district about Kokomo, not far north from Leadville, are very similar in character to those at Leadville. They occur, however, in a somewhat higher division of the Carboniferous, and the ores as a rule are not so decomposed and oxidized, and the transition from the original sulphide character of the deposits to the oxidized condition is more easily shown.

RED CLIFF GOLD DEPOSITS.

At Red Cliff, still further north of Leadville, in the Valley of the Eagle, the same geologic series are found, penetrated, as at Leadville and Kokomo, by eruptive sheets. In the limestones at contact with the porphyries, much the same classes of ore deposits occur, but the peculiar and instructive feature of the camp is the rich deposits of gold in chambers and cavities in the hard and usually unproductive Cambrian quartzites resting on the granite.

The gold in these chambers often occurs as nuggets. The quartzites dip about 10° N. E., and between their bedding planes lies the ore. The so-called contact or bedding plane between one stratum of quartzite and another is clearly defined. At this line there is a filling so to speak of " brecciated," broken up quartzite fragments cemented by iron rust and at times by iron pyrites. The thickness of this breccia varies between four and six feet. Ore chimneys on this breccia occur at intervals.

Their presence is indicated on the outcrop by seams of rusty clay, which lies on top of the ore body and follows it along the roof of the deposit for 100 to 200 feet, then thins out gradually and disappears entirely; at the point of its disappearance, unaltered iron pyrites set in.

These ore chimneys are about 4 feet in width, their thickness is limited to the space between the floor and roof. The quartzite roof is always smooth, but the lower quartzite floor is rough and corrugated and shows chemical action on it attendant on deposition of ore. The floor at times is impregnated with ore which does not, however, extend any great distance into it. Though the ore chimneys are from 4 to 6 feet wide the pay ore is only a few inches, swelling from floor to roof. The pay ore in the oxidized rusty portion yields 7 ounces gold and 50 ounces silver.

In mining, the floor is followed as a guide. Individual ore chimneys are connected laterally by ore chutes like a network. These ore chimneys divide and separate, the branches reuniting or again splitting up. The whole ramification comes together again at intervals in one main chimney. The rock filling the space where the divergence has taken place is the same as the breccia filling, only more compact and impregnated with pyrite. These fillings are left standing as pillars after the ore is mined.

To sum up, the characteristics of these deposits are :

First. The outcrop of the ore chimney indicated by what is locally called a "joint-clay."

Second. A zone of oxidation for 200 feet, which gradually merges, as the natural water level is approached, through a zone of mixed oxides and sulphides to the zone of unaffected sulphides.

Third. The "joint-clay" gradually disappears as the sulphides are approached. The ore on analysis shows sesquioxide and sesquisulphate of iron, silica and alumina and sulphate of barium.

In the Ground Hog mine the ore chimneys are 600 feet apart but are probably connected. They abound in nuggets; the latter are sometimes twisted like bent horns ; in other chutes they are lumpy, composed of crystalline gold particles cemented together by sesquisulphate of iron and horn silver.

Nuggets are found in troughs in the quartzite floor imbedded in clay associated with rich silver or horn silver ore. With the nuggets are lumps of sesquisulphate of iron carrying much gold. This proves, according to Mr. Guiterman, that the secondary deposition of gold in crystals was through the medium of persulphate of iron derived from slow oxidation of iron pyrites, and is an admirable confirmation of the theory as stated by Prof. Le Conte in his Geology.

ASPEN ORE DEPOSITS.

The Aspen mining region is geologically related to that of Leadville; each is on the shore line of the old Archæan island of the Sawatch, one on the east, the other on the west, opposite one another, but about 50 miles apart.

The ore deposits occur in the same general horizon, viz., the Lower Carboniferous.

Both regions show intense disturbance, both by volcanic intrusions of igneous rock, folding, and faulting. The process of ore deposition in both regions has been an actual replacement of the country rock by vein material.

At Aspen the ore is not found in *actual* contact with the overlying eruptive igneous rock, but at some depth down in the limestone, at a zone where the "blue limestone" becomes dolomized, or as Aspen miners say "passes from blue lime into short lime."

The mines of Aspen are situated in Paleozoic strata reclining upon the slope of a narrow ridged mountain, forming a granite spur "*en echelon*" with the Sawatch range.

The strip of country in the vicinity of Aspen constitutes the dividing line between the two distinct uplifts of the Sawatch range on the east, and the Elk mountains on the west, and has been successively affected by each upheaval.

The Sawatch upheaval was a gradual elevation of this mountain mass resulting from a gradual subsidence of the adjoining sea bottoms, which caused the sedimentary beds deposited in those sea bottoms to slope up at varying angles all along the ancient shore line toward the central mass of the Archæan island.

The Elk Mountain range, which extends to the west and south of this region, was upheaved later than the Sawatch, with greater violence and eruptive energy, and the upheaval was accompanied by enormous intrusions of eruptive rock which were forced into the sedimentary strata already shattered by the forces of upheaval, in great "laccolites," or solid masses, and spread out through them in every direction in the form of dykes and intrusive sheets. The surface exposures of these igneous bodies cover areas of twenty-five to thirty square miles, and their extension below the surface is doubtless very much greater.

PLATE. LXXXVII.
Aspen Mountain.

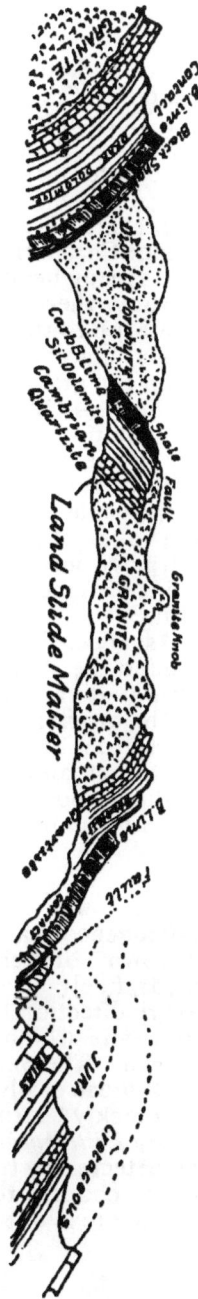

167

The intrusion of such enormous masses of foreign matter must not only have greatly disturbed the beds within the region of upheaval, but also have so expanded the volume of the earth's crust in this area as to cause a severe lateral pressure in the adjoining region. That adjoining region was Aspen and its neighborhood.

It would be just in the strip of sedimentary beds along the Aspen Mountain ridge, which is backed by a projecting point of the unyielding Sawatch Archæan, that this compression would be most severely felt, the Sawatch granite mass acting as a point of resistance against the intense lateral compression caused by the younger Elk Mountain uplift.

The sedimentary beds resting against the Archæan correspond generally, with slight differences, to those in the South Park and Leadville region in a similar position.

The latter were deposited in a partially enclosed bay, now constituting the South Park basin, the former on the west side of the Archæan island in a wider and deeper sea, and on this western slope the beds are generally much thicker than those of corresponding geological horizons on the east.

STRATIGRAPHY OF ASPEN.

1. The horizons represented are the Upper Cambrian quartzites, 200 feet, resting on the Archæan granite.

2. Silurian silicious limestones and quartzites, 340 feet.

3. Darker limestones, rusty

PLATE LXXXVIII.

Section of Aspen Mountain.

brown and dolomitic at base, blue compact and pure on top, 240 feet. (These are Lower Carboniferous.)

4. Carboniferous clays and shales and thin bedded limestones, 425 feet. These belong to the Weber grits (Middle Carboniferous).

5. A series of variegated green and red sandstones, clays and shales, some limestones and red sandstones of the Upper Carboniferous.

6. Heavy bedded red sandstones (Triassic).

Above these again are several thousand feet of Cretaceous strata, up to the base of the Laramie coal beds. (The Cretaceous, however, and the Jurassic do not rest immediately upon the granite).

Diorite.—On Aspen Mountain is a bed of "white porphyry" (diorite) in the black shales, 60 to 100 feet above the top of the blue limestone. It is 260 feet thick on the slope back of town, but thickens considerably to the south, and is traceable to Ashcroft. It appears to extend also across the valley of Roaring Fork to Smuggler Mountain. Small intrusive sheets also occur in the lower quartzites near the point of Aspen Mountain and on the east face of Richmond Hill.

As affected by the Sawatch upheaval, these beds wrap around the Archæan mass, resting against or dipping away from it at varying angles.

The quartzites and limestones cross the valley of Roaring Fork from Smuggler Mountain to Aspen Mountain, striking northeast and southwest, dipping northwest. The angle of dip is about 45°, varying from a minimum of 30° to a maximum of 60° in "flats" and "steeps."

THE ORE BODIES.

The lower carboniferous "blue limestone" is compact, homogeneous and composed of pure carbonate of lime. The "brown" or "short" dolomitic limestone is of a dark gray color, finely crystalline, finely granulated and traversed in every direction by a network of minute veinlets containing iron salts, which, when oxidized, color the surface a rusty brown. The oxidation along these minute veins makes the rock break easily into dice-shaped fragments giving the rock a "crackly" structure, hence its local name of short lime.

Ore Distribution.—The outlines of the ore bodies cannot be detected by the eye, owing to the gradual transition from ore to country rock.

The ore is not confined to the brown dolomite below the

so-called contact, but several ore bodies extend 20 or 30 feet above this contact into the blue limestone and in some cases follow the lines of cross-fracture entirely across the blue limestone.

The ore is not confined, either, to a definite plane or contact between two dissimilar beds of limestone and dolomite from which its solutions have eaten into the underlying dolomite, for in the first place there is not one single contact, but many; and if this so-called contact constitutes an essential condition of ore deposition, there is no reason why it should be confined to the one and not found in the others where the rocks have the same composition. Again, ore-bearing solutions would not be likely to eat upwards for any great distance from the contact plane if they entered the beds along this plane.

This so-called contact plane is well defined on Spar Ridge and continues down with the dip in the underground working, but ore bodies occur above and below it.

The rock thus mineralized is dolomite in most cases, but it is none the less above the true bedding plane called the contact.

In other parts there has been fracturing across the beds as shown by a vertical breccia of limestone fragments with a cement of iron oxide and manganese.

Over the ore bodies are lines of open cavities following the lines of cross-fracture, through which the ore solutions passed which deposited the ore bodies. These caves are now being hollowed out by water descending from the surface dissolving the limestone in the roof and flowing off along the floor, depositing a mud of silica, alumina, lime, magnesia and iron oxide.

Hence this contact is not necessarily the only ore channel of the district, and other channels may be sought for.

Portions of the ore bodies have been formed by solutions percolating through cross-fractures and spreading out between the parallel bedding planes.

This would happen if these solutions derived their metals from the overlying porphyry, for it is separated from the limestone by argillaceous shales which would be impervious unless fractured across the bedding. The analysis of the lime mud at bottom of the cave shows by its preponderance of alkalies, which do not exist in the composition of either brown or blue limestone, that the waters dissolving it came from the porphyry. The waters brought both alkalies and silica from the porphyry, and probably the iron and baryta.

DOLOMITIZATION.

This is a secondary process upon the blue limestone by magnesian waters, which is proved by irregular tongues of dolomite extending up, into and across the blue limestone. The lenticular bodies in the Durant cliff point to the same fact. The crackly structure of the brown lime results from the replacement of a molecule of lime by a molecule of magnesia, involving also a contraction in volume of the rock itself, which would cause it to separate in angular fragments, the intersections filled by material more soluble than the rock itself.

The magnesian waters may have been connected with those which brought in the vein materials.

In the ore bodies the partially mineralized rock on the borders of the ore is changed to dolomite, hence dolomitization either preceded or accompanied ore deposition.

Mr. Emmons suggests as *probabilities* only, that the porphyry intrusion preceded the faulting;

That the ore deposit followed the intrusion of porphyry and also the principal faulting movements;

That small movements have taken place in recent times both in the strata and contained ore bodies since the oxidation of the latter; that at the time of the great faulting, the beds may not have attained entirely their present position.

In the vicinity of Aspen Mountain ore bodies, the strata appear to have been synclinally folded and faulted between the main Archæan area on the east, and a mass of granite at the western extremity of the mountain, thus producing a second series of oppositely inclined beds, also containing a few ore bodies. Intrusions of altered eruptive diorite occupy a prominent position in the intervening trough and may have seriously faulted or dislocated the strata in the depths. The bulk of the Aspen ores are largely oxidation products of argentiferous minerals with true silver minerals, associated with calcspar and baryta; it is a "dry ore" requiring to be mixed with silicious lead ores before it can be treated. Such rich ores as polybasite and brittle silver occur also.

A great deal of the ore consists of fine grained steel galena, very rich in silver.

ASPEN AS A PROSPECTING GROUND.

Aspen again is an example of a region that had often been *skimmed* over by the prospector and abandoned before the final thorough prospecting revealed its great riches. Years

ago some prospectors found signs of " float " and " blossom "
cropping out under the blue limestone of Spar Ridge. They
even went so far as to sink an incline of a hundred feet or
more, but though they found ore, its character was so low
grade, that the mine was for a long time shut down and
practically abandoned. Then an enterprising individual conceived the idea of boring down on the sloping back of the
limestone in the adjoining Vallejo gulch, to tap the ore
body, already discovered along the outcropping, on the
underside of the limestone. At about 150 feet deep the
limestone was pierced, and an enormously large and rich ore
body was discovered. Immediately, the original locators
began again with all speed to push on their incline, and
then originated the celebrated " apex and side line " lawsuit.
The original locators had the apex on the outcrop. They
therefore claimed the whole mountain, and tried to drive
out the side line men. Finally a compromise was effected,
but that boring down on the back of the limestone and its
discoveries led immediately to an army of prospectors examining the mountain, and it was astonishing how many ore
deposits were discovered in a region that was supposed to
have been prospected and given up as no good. Of course
Aspen is an example of "*richness with depth.*" A dangerous
precedent and encouragement to that often ruinous policy
of running long cross-cut tunnels to cut an ore body at
depth, which has only proved indifferently good near the
surface, on the fallacy we have before alluded to, of the
improbable probability of "richness increasing with depth."

AN EXAMPLE OF PROSPECTING.

Now supposing our prospector was the first man to enter
that region years ago. What signs were there to lead him to
think it was a good prospecting ground ? Supposing him to
be fairly versed in geology, he would have noticed, as he
came down over the Sawatch range, that the Paleozoic
strata he had observed as ore-bearing at Leadville, out-
cropped also on this western side, together with the " blue
limestone ; " secondly, he would have noticed the presence
of large masses of *eruptive* rock constituting the Elk range ;
thirdly, he would observe the region was much disturbed,
that the strata were intensely folded, and intensely faulted.
All these signs he would have considered likely. Then
after following up the various creeks, he would select such
spots as where he saw the massive blue limestone out-

cropping. He would readily find this bed from its relation to the granite and Cambrian quartzite below. He would look for places where porphyry was intruded into the limestone or where great masses of it lay above or in vicinity of the limestone. This would probably have led him, on nearing Aspen Mountain to give that mountain more than a passing look. He would notice that the strata on Aspen Mountain were very much disturbed and faulted, that a spur of granite, quite out of place, came right up through the middle of the mountain, that strata were pitching in various directions off from this, and moreover that in the lap of this fault-fold was a very thick bed of porphyry. He would observe the line of change from the blue limestone to the dolomite, and at that line he would have prospected and found and followed up the " blossom " at the line, consisting of calcite and baryta running in a rusty line, like the outcrop of a coal seam, all up the side of Spar Gulch and so he would have discovered the great Aspen ore-deposits, and by following up the indications along the outcrop and locating claim after claim as along an outcropping coal seam, he could have secured practically the whole " apex " of the hill, and become master of the mountain and all it contained; but had he known then the litigation of "side line and apex" that was to arise, he should have gone further, and located claims covering the side line, on the back of the sloping limestone ridge, leading down into Vallejo gulch. But again he might, like the original first discoverers, have become disheartened with his find on testing the outcropping ore by assay or mill run, and finding it so low grade near the surface. On general principles in this respect he would have been right.

Now having thoroughly explored the little Aspen Mountain, he would observe that much the same formations crossed the creek and entered into Smuggler Mountain, though much obscured by heavy glacial drift. Here he might have located fresh claims on this hill, and become the owner of the celebrated Smuggler, Regent and other mines with their untold wealth. Thence he might have continued his successful trip, and followed the same so-called "contact" outcrop for miles on to Ashcroft. It must be remembered here, however, that in locating all these claims, whilst the prospector may drive his location stake at every 1,500 feet, he is required within sixty days after location, to dig a ten foot hole in each location. As this may be a little difficult for him to do he generally enlists others in his enterprise, to assist him, and enters some of the claims in their names, to prevent the discoveries being jumped by a horde of prospectors who press in as soon as anything is found. Good advice to a prospector, is to keep very still and "mum" about his discoveries until he has well secured them, and to be very careful how he " opens his head" to any one. Commonly a prospector who has "struck it," comes into town, fills up with whiskey, " blows it in,"and then, " blows it off " all over town about his discovery, and is elated to find himself the hero of the hour. The result is, before daylight the following morning, a hundred men are chasing one another in the direction of his discovery, and before a day or more is over, the mountain is covered with locations as close as graves in a city churchyard, and in a week's time these locations are covered again by a second layer, as the saying is, " several feet deep."

A boom follows. The offscourings of the country pour in with the saloon, dance hall and gambling hell element. A murder or two follows. Lynch law takes a hand. Then a horde of real estate men come in, and lots are sold at fabulous prices, and the town is inflated with a population and everything else usually far above the capacity of the mines to support. A collapse follows, and .a steady retreat of hollow-eyed, disappointed adventurers. In time the town and camp assume their lawful proportions and business settles down to its lawful regime.

Whilst all this has been going on, and amidst all the fuss and bustle and " hooraying" of real estate " boomers " and so forth, some prospectors have been quietly trying to follow up the first desirable indications into the neighboring region, resulting often in an extension of the ore-bearing region. Some of these locations are " bona fide " and valuable. Other " holes in the ground " are dug on the merest pretext of indications to catch the ignorant, adventurous tenderfoot capitalist purchasers, or " suckers." An investor going into the camp at such a time, finds a fabulous price placed on every prospect, whether genuine or false. As a prudent man, he either beats down such prices, or concludes he will visit the camp a little later, when the excitement and inflation has gone down, and when things are on more of a business footing, and something like the real value of the camp has been found and proved. Of course in such a gambling speculation, by such prudence and delay he may lose a lucky chance, but he has preserved his prudence and escaped being wofully bitten.

Perhaps in a month's time, the discoveries are found to be merely superficial, the boom utterly collapses, and the dreary sight is seen a little later, of a desolate village, with frame houses and log cabins, and possibly a mill or two, for mills are sure to follow, lying in wreck and ruin, a home for the owls and the bats; or else the genuine discovery produces one or two mines and supports a handful of population legitimately.

Again, a region like Aspen may disclose a limited number of very rich ore deposits, but sufficient to support and sustain a fair sized town.

But the most important and most lasting discoveries of all are of areas producing an immense quantity of low grade ore such as Leadville. This gives an opportunity for a great number of mines and for the support of a large and permanent town.

CHAPTER XIII.

EXAMINING AND SAMPLING MINING PROPERTIES, PROSPECTS OR MINES.

A prospector may be, or become, a "mining expert" and be called upon to make examinations of mining properties, whether prospects or developed mines, so a few suggestions may prove useful.

Mining properties of the precious metals are generally of two kinds, those containing ore deposits in place, such as fissure veins and blanket deposits and placers, the latter being gold-bearing. In both cases, and especially in the former, the character, position and other relations of properties are infinitely varied, so that no hard and fast rule can be given to suit all cases; certain rules, however, will generally apply.

A mining engineer receives a letter from a company telling him to go to such and such a country or region and examine and report on a certain property that has been offered them. Such a mandate is usually accompanied by a letter or report from the owner or parties offering the property, giving the owner's description of the same, or else the report of some expert on it. As a general rule such reports give the most favorable view, and in some cases must be taken "*cum grano salis.*" To the mining engineer they give some sort of an idea as to what the property may be like. As to its value, etc., that he proposes to find out for himself. The company sometimes asks him to examine with a view to verifying or modifying or contradicting such reports.

The region, the country, the character of the deposits, the local conditions, may in all probability be comparatively new or strange to him. Prior to starting he may make inquiries in mining circles if anything is known about the region or district. If there are any published mining or geological reports or maps, he will consult these. Finally he starts out with as little baggage as possible, usually a small hand bag, containing a few necessaries ; a tapeline, geological pick, clinometer and compass and note or sketch book. His dress is generally a suit of corduroys, leather gaiters and strong boots.

As he enters the region by rail or on horseback he notices the main geological features, whether the rocks are granitic, sedimentary, or eruptive. Finally he reaches the camp, calls on the owner or superintendent and rides up with him to visit the mine. If he should "lay over" for the afternoon in the village, he may as indirectly as possible try to pick up any gossip there may be afloat relating to the property. He is at once impressed with the accessibility or inaccessibility of the property, and estimates the probable cost of bringing down the ore to the mill or to the railway track, and observes the proximity or absence of timber and water power. At last he reaches the mine, dines at the boarding house, and is then taken over the premises by the superintendent. His first attention is directed to the surface character of the property, its topography, whether rolling, smooth or precipitous, whether it is high above the valley or near down to it, whether the mine is high or low as regards the water level or drainage system of the neighborhood, whether the property is conveniently situated for working the mine and transporting the ore, etc.

Accessibility is an important matter. In some regions, such as in the San Juan district (Colorado) for example, mines and prospect holes are sometimes on the top or sides of mountains or precipices, thousands of feet above the valley below, located at spots one would think only an eagle could reach ; prospect tunnels, too, are driven where there appears scarcely a foothold for a squirrel. No spot, however, seems too inaccessible for the prospector. At a glance the

176

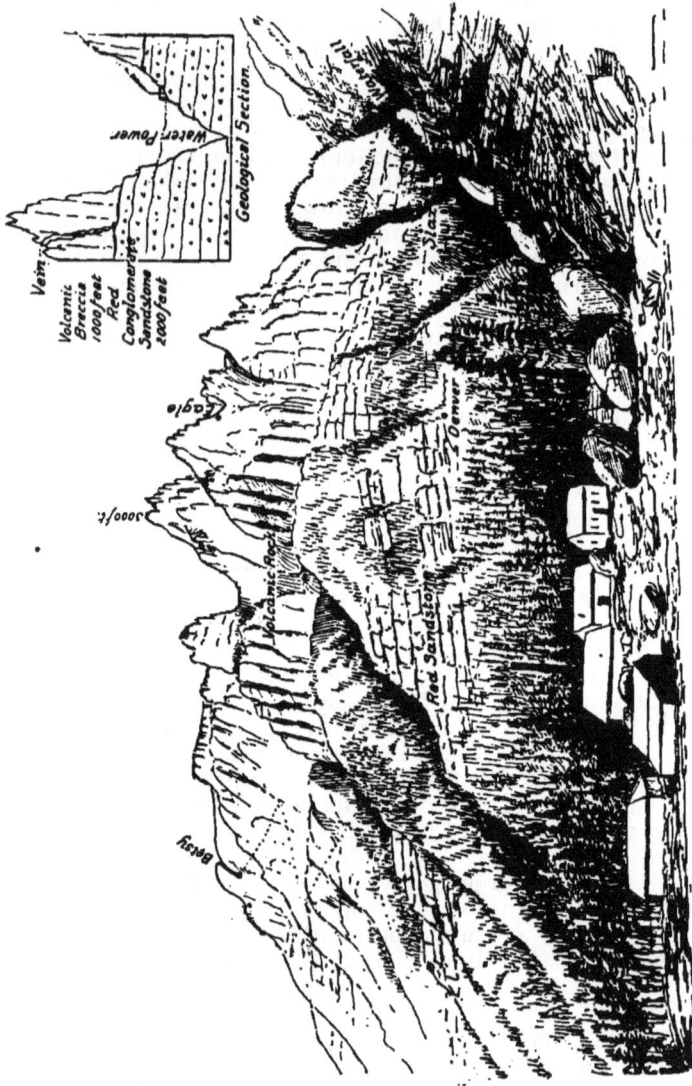

PLATE LXXXIX.

A Somewhat Inaccessible Gold Property (San Juan).

engineer sees that in a property situated in such a region, accessibility is one of the first and often most formidable problems to be considered.

To some of these mines are long zigzag trails cut in the side of the mountain. The engineer calculates how much the owners of the donkey or "burro" train will charge to bring that ore down to the valley or mill. He argues that a mine at that almost inaccessible height, ought to carry a good deal of pretty high grade ore to pay even for transportation by the "burros," let alone the cost of freight afterwards to distant smelting works. On the other hand a mine whose workings open out within easy access of the valley or railway track, could afford to carry less valuable ore. Then there is timber and water power to be considered, the former for timbering the workings of the mine, the latter for running a stamp mill, or for supplying steam power to the engines of the mine. If there be no water power, and the vein carries free gold, the ore must be carried down to the nearest stamp mill. In a young or "virgin" property or prospect, the engineer will look out for a convenient site for such a mill, under a developed property; if there be a mill on the premises, he will examine and report on its capacity and suitability for treating the ores. A mill site must, of course, be selected close to some water power. In some districts there is a superabundance of water, in others a serious lack of it, or the supply is meagre at certain seasons, or is frozen up in winter. Some mines are quite dry, but generally they will supply enough mine-water from their workings to afford steam power.

He observes the character and dip and direction of the veins if exposed on the surface, examines any prospect holes on them, and takes a few samples for assay. He will consider the nature of the ore deposits, whether they are in fissure veins in crystalline rocks, or blanket deposits in sedimentary rocks, or contacts at junction with eruptive porphyries. A great variety of local and minor details have to be noticed which can hardly be specified.

Having looked over the surface, he enters the tunnel or workings with the superintendent. As he passes along, the latter is likely to call his attention to this or that spot, as especially good, and naturally he rather overlooks the poorer parts of the mine. He may suggest the advisability of taking samples from such favorable spots. The engineer, however, takes little heed, as, if he were to confine his attention to, and sample only these choice portions, he would

obtain an incorrect estimate of the average run of the mine.
Moreover in some cases it is well to be on the look out, lest

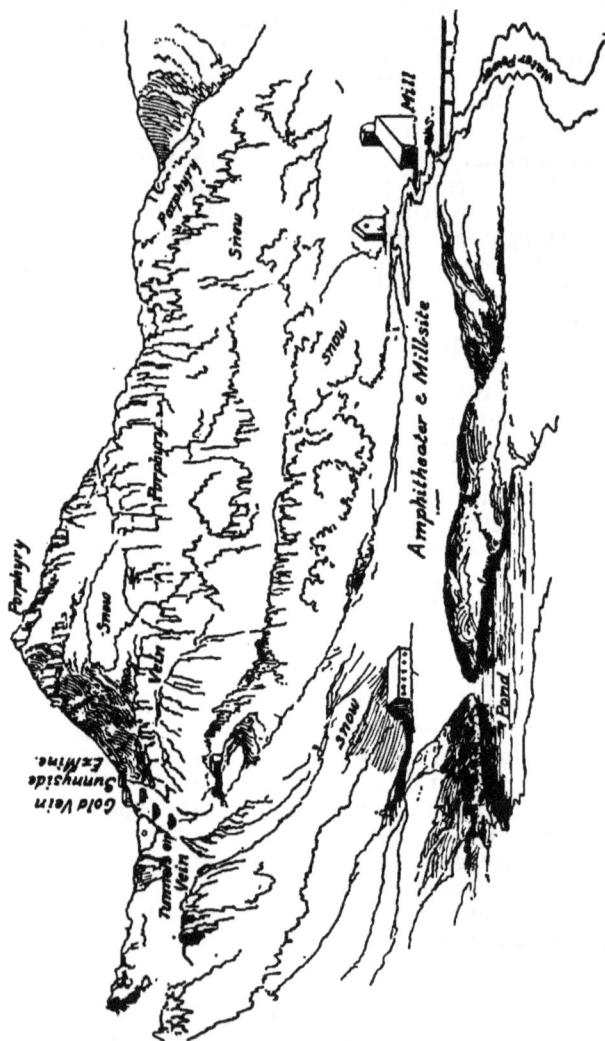

PLATE XC.

Sunnyside Extension Gold Mine, San Juan. Example of a Rough Hasty Field Sketch of a Mining Property.

these points to which special attention is called by the miner,
be previously "salted" or "put up" for the expert, and
charged in various ways by rich ore.

Having traversed the workings and obtained a general
idea of the position of the vein and ore bodies, and taking
an inventory of the amount of development, length of drifts,
shafts, etc. (the latter he can obtain from a map of the mine
in the superintendent's office, a copy of which he will send
to his company) he asks the superintendent to leave him and
his assistant and vacate the mine, as he does not wish any
one except his assistant to be with him whilst he is taking
samples for assay.

SAMPLING.

Now begins his hard and most telling work, the time and
labor depending very much upon the size and amount of
development in the mine, or the degree of accuracy necessary. Taking a large strip of muslin, he cuts part of it up
into small pieces about the size of a pocket handkerchief,
these are to contain his samples for assay, when quartered.
Then he takes the remainder of the muslin, or better still an
ordinary candlebox, this to catch the mass of small fragments he detaches from the vein with his pick.

Now with a light pick or with a chisel and hammer he
begins, either from the entrance or end of the tunnel, to
detach small portions of the rock, cutting a rough groove
across the vein. Sometimes the tunnel occupies the whole
width of the vein in which case he will have to make a circular groove clear around the tunnel, across floor, roof and
walls as shown in Plate XCIV; the fragments from his work
drop into the candlebox or onto the muslin.

According to the length of the workings or the need for
great accuracy, he repeats this operation at intervals which
may be every 5 feet, 10 feet or 20 feet ; at intervals of say 20
feet, he masses and mixes together all the samples, breaks
them up as fine as he can on a shovel and divides the result
into four parts, throws away three parts and retains one.
This he reduces to a fine powder and wraps up in the small
muslin pieces, ties it up and seals with sealing wax, marking
on it the number and other notes, such as 10 feet from
entrance, etc., with an indelible pencil. This work he continues till he has reached the end of the tunnel, which if it
be 100 feet long will give him from 20 feet intervals, five little
sacks of powdered ore for assaying. As a check upon this
work and for reference in case of any accident to or any
tampering with his samples in transit, he will occasionally
take a "grab" sample from his broken rock before quartering it. Here and there, too, he may take a chunk of some

180

peculiar rock such as a porphyry or some peculiar streak in the gangue, these he will put in his coat pocket and keep on his person.

It is sometimes important in a vein to find out what rock or portion of rock carries t h e most value. For instance, on one occasion we examined a vein said to carry gold clear across its e n t i r e width of some 50 feet. Now this so-called vein proved to be a decomposed dyke of porphyry impregnated w i t h pyrites and free gold, a n d through the dyke ran a net-work of little narrow quartz veins or veinlets. On sampling, whilst we took samples clear across the whole width of the vein, we kept those fragments which came from the quartz veinlets apart from those which came from the porphyry

PLATE XCI.

Natural Appearance of Mine on a Blanket Ore Deposit.

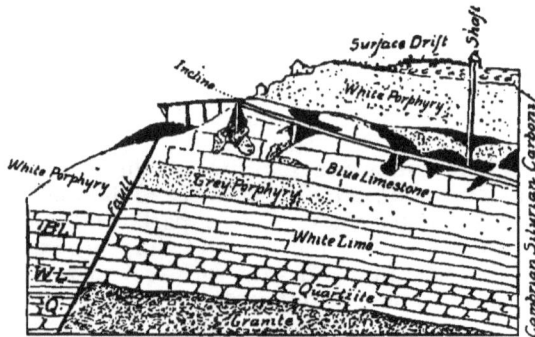

PLATE XCII.

Geological Section Showing Workings and Ore Bodies in Contact Blanket Ore Body. Shaded Portions are All Worked Out.

gangue. The result was we found the porphyry, consti-tuting of course the main element, to be barren and the gold to be concentrated in the quartz veinlets constituting a minimum of the width.

So in a vein there will generally be parts richer than others, " pay streaks " as they are called, which it is important to distinguish, also certain metallic minerals in the vein carrying greater values than others. Thus the pyrites, if undecomposed, may prove too poor to treat for gold, or in a silver mine, streaks of gray copper may be very rich, whilst bodies of coarse galena may be very poor. In a gold mine it is important for the engineer, if he can, to find out to what depth surface decomposition or oxidation has penetrated, because in this brown rusty matter will likely be most of the "free gold;" whilst when the unoxidized pyrites makes its appearance the ore is no longer free ore, but must be treated by some process other than that of a stamp mill,

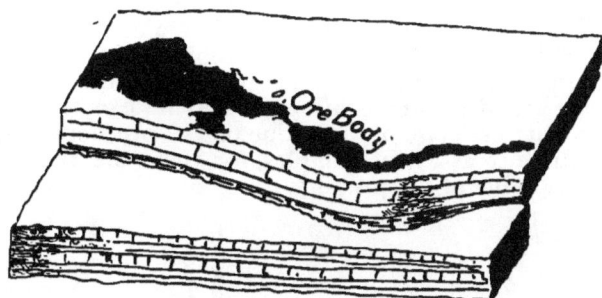

PLATE XCIII.
Plano-Section of a Flat Ore Body.

and with the incoming of pyrite the palmy days of the gold mine may be at an end. Sometimes, however, though the oxidized brown gossan may play out and succeed to white quartz, the latter, if it be not too hard, white and " hungry," may still continue to carry free gold in it. Again in the veins, with their descent into depths, greater or less richness may occur or different varieties of ore set in, or absolutely barren quartz, so if there be shafts or tunnels driven on the vein a distinction should be noted with descent as to values found at different levels, also as to character and richness of the ore above and below water line; the latter corresponds to the average drainage level of the country.

This completes his underground examination. Whilst in the mine he may make a rough sketch or two of the vein

182

showing the general disposition of the ore bodies or any peculiarities. On emerging and carefully securing his samples beyond reach of their being tampered with, he selects a convenient point, possibly on a neighboring hill facing the property, and takes a general sketch of the property in pencil or water-colors (see Plates LXXXIX, XC, XCI and XCII), also makes a pencil sketch and ideal section of the hill, showing the position of the vein and its workings (see Plates XCV and XCVI), the amount of ore stoped out and the amount presumably in place intact; to estimate the latter is often a difficult and uncertain problem. He may make some sort of estimate as to the reasonableness or not of the price asked and give his estimate; he can form, however, no true estimate of the value of the ore bodies till he has had time to assay his samples, for these are the crucial test of the value of the property.

PLATE XCIV.
Expert Taking Samples.

PLATE XCV.
Fissure Vein Outcrop on Hillside Showing Surface Workings.

In writing up his report at his leisure, which will most likely be read at a general meeting of the company, he cannot be too clear, simple and explanatory in his account and its details, as it is to be remembered that the company is likely largely to be composed of men unacquainted with mining and mining terms; he must therefore not take it for granted that they know what "stopes," "adits" and "gouge," etc., are, but explain as he goes along, accompanying his remarks with rough sketches to make his meaning clear and put the members of the company as much on the ground as possible. We ourselves have found that it is not necessary, generally to make elaborate notes in the field or to write pages of reading matter then, provided we make many sketches and on them put down items such as length of workings, etc., etc. The sketch is generally the notes, and

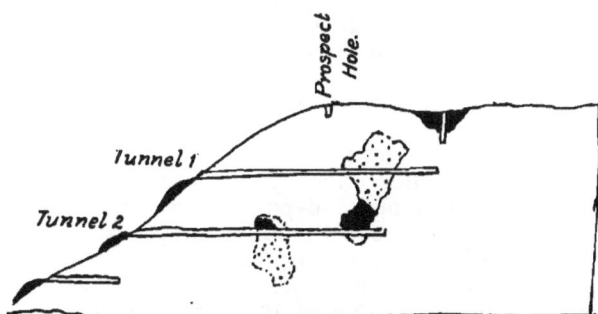

PLATE XCVI.

Cross Section of Vein Showing Workings. Dotted Portion is Ore Body, Shaded is Ore Worked Out.

when the engineer returns home, his sketches will recall vividly all he has seen and from these he will write his report. Upon certain matters, however, such as involve numbers, he should be very accurate in writing notes and not trust to treacherous memory for them.

DESCRIPTION OF PLATES.

In Plate LXXXIX we have an actual example of a rather inaccessible property in the San Juan, which with Plate XC shows the kind of sketches to accompany a report. In Plate LXXXIX with its section, it will be observed how very high up the mining holes and prospects are perched, in most cases over 1,000 feet above the valley, and again, before the ore can be brought over to the mill, a ravine of a hundred feet deep occupied by a boiling torrent has to be crossed. Some of the properties might be worked perhaps by a suspension tramway thrown across the gulch. In another case a trail has had to be cut in long z gzags of some miles before the bottom of that could be reached. Another disadvantageous feature in this property is the number of scattered veins, none of them very rich by itself; this involves a separate plant or workings for each. One good vein would be better than all these put together. There is fine water power on the property and plenty of timber.

In Plate XC there is one fine rich gold vein easily accessible and easily worked ; below it lies a natural basin and abundant water which makes it an admirable location for the stamp mill. This Plate gives an idea of the rough kind of a sketch the expert makes on the ground, which he embellishes and elaborates on his return.

Plates XCI, XCII and XCIII show :

(XCI.) The surface appearance of a "flat" or contact blanket on the side of a hill, such as at Leadville.

(XCII.) Cross-section showing the position of ore bodies, the portions worked out and portions probably left in reserve, also the workings of the mine and the geological section, together with a prominent fault.

(XCIII.) Is a somewhat ideal sketch of the probable relations of a flat ore body if the surface matter were removed, or rather if it were opened like a book.

Plate XCIV shows the expert taking samples in a tunnel driven in the vein ; the vein in this instance, being a very large one, occupies the whole width of the tunnel ; this is not generally the case, the vein and ore body are more commonly observed about the middle of the roof, *i. e.*, if the vein is small.

Plate XCV shows the outside appearance of a fissure vein with three tunnels down in it, and Plate XCVI shows cross-section and profile showing the tunnel and the ore bodies so far discovered in the quartz gangue and how much has been worked out.

CHAPTER XIV.

SALTING MINES.

In these days when, owing to the depression of silver, so much attention is being turned towards gold and gold mines, too much care cannot be taken by those investing or act ing as examiners or experts in gold mines, that there are no tricks played upon them by the astute miner; for "for ways that are dark and tricks that are vain" the western miner is at times "peculiar." One of these tricks is what is known as "salting" mines or ledges ; that is, by various means and ways introducing into the mine or into the samples taken from it, certain rich minerals which do not rightly belong by nature in the mine or property, in order to raise the value of the mine in the eyes of the investor or expert. When samples are taken from such a tampered-with mine the values and results must be accepted *cum grano salis*, with a very large grain of salt indeed. Whether this classical allusion be the origin of the word "salting" we do not know.

"Take care you ain't salted" is the advice to the inexperienced investor or novice expert. So clever are the miners, that cases are on record where even a most experienced expert has been taken in, and comparatively, or wholly valueless properties sold for large sums, the purchase followed later by woeful dismay and surprise, when dividends were called for and did not appear.

Gold mines of all others, are the most easy to salt, hence the precaution in these days is timely.

Whilst a mining engineer or expert can hardly prevent salting, with care he can, and ought to be able to avoid being taken in ; to be forewarned is to be forearmed.

On entering a mining camp in the far West, especially in the more remote outlandish districts, an investor or an expert, may consider that the whole village, from the ho el bell-boy to the mayor, (who, by the way may be the principal saloon keeper) is in league against him. Directly he arrives, everybody in town wants to know his business ; on this he should keep as mum as possible, and, if he can, throw impertinent inquirers off the scent. The idea is, "Here is a capitalist to fleece and an expert to delude." Every one, too, has a "hole in the ground" of his own to present. Should they get wind of the particular property in view, there are confederates and middlemen anxious to share the spoils. Moreover, it is considered to the general credit of the camp to sell a mine, be it whose it may, good or bad, and if you mention any property, you will invariably hear it "cracked up." The eastern "tenderfoot" is somewhat of a "sheep among wolves" in such a camp. The expert, too, is at a certain disadvantage on entering into a strange mining camp, not being familiar with the local conditions. Ores for instance in one section or region are not always of the same value as similar ores in another, the rocks may look new and strange to him, and there are a hundred local conditions known only to the resident miner. It would be well, when possible, for an expert, before passing a decided opinion on an important property, to stay around in the vicinity for a while till he knows the "hang of things."

On his way to the mine there will be plenty to fill his ears with the untold value of the property he is about to examine, this friendly duty is not unfrequently performed by an officious middleman. To favor and "soften up" the expert's mind and heart and make him "feel good" toward the property, attentions of all kinds are showered on him. He is driven about town like a nabob, and if he shows a weakness for a "wee drappie," champagne and whiskey are at his service *ad lib.*, as judicious preparation for the coming examination. It may be observed here, that attempts are made sometimes to "salt" the expert as well as the mine, not merely by befuddling his brain with intoxicants, but by offering bribes, and as an expert is often not too well off, the latter is a great temptation.

We will now suppose, after this ordeal, he goes to the mine with the superintendent or miner. All may be, and we may say generally is, honest and square, or it *may* not. The expert looks over the exterior and surface signs of the property, studies the outcrop of the vein on the surface, its probable surface continuity, the advantages and disadvantages of the situation of the mine, its proximity to railroads, smelting works, markets, etc., and then enters the mine in company with the miner. As a rule the latter will naturally point out to him the richest portions and ignore the poorer; sometimes he excuses himself from taking him down into the latter because it is dangerous or full of water. If full of water the expert if possible should have it pumped out. He may suggest here and there, that such and such a spot would be a good one for the expert to take his samples and so forth. The expert of course assents to all he is told, but with one eye open, and does not stop to take any samples for assaying until he has seen the whole of the mine, then he requests his companion to go out on the dump and smoke his pipe there, as he insists upon having no one with him in the tunnel when he is taking his samples for assay. He will be inclined to rather avoid those particularly favorable spots suggested to him by the miner, as probably giving too rich an average for the general run of the mine, or as not impossibly being "fixed" for him. If he suspects the latter, he will take a sample or two to see if the mine has been tampered with, taking a little of this out on the dump crushing it and washing it in an iron spoon. If a very astonishing amount of gold colors show up, his suspicions are aroused. The judicious miner does not generally want to salt too heavily, for fear of the enormous results exciting suspicion, but despite his care he nearly always salts a little higher than he intended. In a mine where the rock is hard, a miner may salt by drilling holes and inserting mineral or ore and disguising the hole. In loose ground or one full of cracks, a shot-gun loaded with a moderate discharge of gold-dust will do the work. The skill of the miner in this case lies in his choice of a spot where he thinks it probable the expert will take samples, or in coaxing the expert to take samples from such ground. In hard ground the expert may avoid such salting by having the work blasted out in his presence till a purely fresh, virgin face is shown and then taking his sample. These precautions are not necessary under all circumstances, but only in such cases where the expert has a suspicion that there is an attempt to "put up a job" on him.

After getting his samples, and as many as possible, he will sack and seal them then and there in the mine, and never lose sight of them till he has expressed them to his own home.

Sometimes a mine is so timbered up, that sampling is difficult. Now as they go down the shaft, it may be the expert remarks "I should like to take a sample in this shaft, but it is so timbered up that I don't see how we can do it without ripping out some of these boards." "Why of course, so you oughtter" says the miner, "and see here, I think this board is loose." Now beware lest that board was purposely loosened and behind it the ground is salted.

By taking a great number of samples at comparatively close intervals, provided afterwards the samples are not tampered with, the expert is less liable to be deceived by salting, than

if he took very few. A mine cannot be salted all over from
end to end if it is a large one, only at judicious intervals,
and it will be hard if the expert does not escape some of
those intervals and get some true samples.

Besides taking his regular assay samples by cutting all
around the walls, roof and floor of the tunnels at intervals
of five, ten, or twenty feet, according to circumstances,
crushing, and quartering the debris, and finally sacking and
sealing his sample bags, he should occasionally take a "grab
sample," or a bit of rock at random, or a small sackful
from the great mass of his sample, and put them in his coat
pocket, and keep them on his person, to act as a reference
in case of any possible tampering or accident to his samples
whilst in the vicinity or in transit. He should also take
bulk samples, good sized chunks of uncrushed rock which
should agree with the assay results of his quartered samples.

A disadvantage an expert is under in a strange camp, if
he cannot take his own assistant with him, is, that he is
very much at the mercy of the miner, if any hard work has
to be done, such as blasting or hard digging. Whilst
engaged in such work the miner, if he pleases, has many
chances of scattering around a little gold-dust on the rock
of the vein or the loose dirt of a placer.

Whilst gold-dust is the favorite medium for salting a gold
mine, chloride of gold is sometimes used. The latter, however, is rather a dangerous and barefaced trick to try on
a competent expert, as its quality can readily be detected
by the chemist, it being soluble in water. In a case of this
kind that came to our knowledge, an experienced expert
had examined a certain mine and condemned it. Later, the
owner who was an honorable man, asked him if, as a special
favor, he would re-examine it, as in his absence the assay
values from the mine had of late shown much better results.
The expert reluctantly consented to do this, though contrary to his general rule. In going along the workings he
noticed here and there on the walls, certain patches and
streaks of clay or mud, he had not observed on his first
visit. Guessing what they were, he casually observed to the
miners, "Seems to have been raining in the mine since I
was here." However to the great delight doubtless of the
miners he took several samples of these, and forwarded
them to a reliable chemist. The latter pronounced them
chloride of gold. This of course gave the salting scheme
away as chloride of gold does not occur free in nature, much
less in a mine. The owner of the mine was exceedingly

angry when he learned what the miners had done without his knowledge or connivance. The men themselves being commonly more or less interested in the sale of a mine, are apt to try and salt it without any connivance of the owner or superintendent. We heard of a case in the San Juan district where a mine that was fairly good was about to be examined. This mine carried occasionally specimens of the very rich ore, called ruby silver. Not satisfied with the fair, natural richness of the mine, the miners must needs import into the hole, quantities of ruby collected from other mines in the district, whose men were of course in sympathy with the scheme and probable sale. This was acting without the knowledge of the owners.

SALTING GOLD PLACERS.

Although a gold placer usually covers a very large area of ground, it is possible to salt it. Usually a miner shows up his placer by opening up pits at convenient intervals, so as to cover the property. Nothing is easier than to salt these pits with gold-dust. Consequently whilst an expert will examine these holes and pan the dirt, he should be on his guard, and insist, where possible, on holes being freshly dug in his presence. Even then he is not safe. Generally in a placer, by the cutting of a stream, sections are shown sometimes from grass roots to bed rock. From such he should take and pan samples at different levels in the exposure, this too, privately and without too much supervision of the interested miner.

SALTING ASSAY SAMPLES.

This may be done in several ways. If the expert is imprudent enough to allow a miner to accompany and assist him in breaking down or crushing samples or panning them, then the infusion of a little gold-dust is easy. Again, after the expert has made up, sacked and duly sealed his samples with wax, should he leave them anywhere within reach of the miners, they are not wholly safe, for the miner may insert the point of a fine syringe containing gold-dust into the bag, or he may make a bread mould of the wax seal, open the sacks, and either change the ore for richer, or infuse some gold-dust. Changing of samples for others is not an uncommon trick. The expert cannot watch his samples too closely. He should sack and seal them on the ground, sleep with them under his pillow if need be at night, yet even then cases have been known when the wary miner has succeeded in extracting and changing them for bags, to all appearance exactly similar. The samples are never safe till boxed up and expressed and on the way to the city address. He should never fail, as we have said, to have partial duplicates of these about his person.

If the expert wishes to assay the ore at a friendly assay office near the mine, whilst he is grinding down his sample to dust, an innocent looking miner may loaf in, and whilst watching the operation, accidently upset the ashes in his pipe over the sample. Probably these ashes contain gold-dust, and we might here observe that a single grain of gold smaller than a pin's head may materially alter the results of an assay.

Some years ago an individual who had succeeded in booming a certain placer district and getting up an excitement and a rush, constituted himself as a referee, and professor ; and when miners brought samples for his inspection, the were always found to be very rich in gold. But simila samples from the same spot if uninspected were somehow invariably barren. The wizard's mere look seemed to change the sand into gold, until it was found that he concealed in his finger nails "which were taper" not wax, but fine particles of gold. Hence Midas-like whatever he touched he turned into gold. Whilst the salter may lay traps for the expert, the expert may sometimes lay traps for the salter. An expert, who had reasons to suspect a certain mine he was examining had been tampered with and guessing there was a likelihood of an attempt on his samples, after securing himself with duplicates, left his samples exposed on the floor of his room at the hotel, then went out and hired a reliable Mexican boy to watch his room and report to him immediately if he saw any one enter it. He had not long to wait. At dinner the boy tapped him on the shoulder, and he went to his room and caught the miner in the act of tampering with his samples.

Sometimes miners, if wealthy enough, will go to great expense to salt a property. Some miners took a couple of well-to-do eastern capitalists to a certain placer, panned the gravel before their eyes, and showed up wondrous colors. The investors having been warned of miners' ways, refused to entirely swallow the bait, but told the boys to go ahead and develop the property, and if at their next visit, it show-ed up as well as the pans did on this occasion, they would

buy it. When the easterners were gone, at a cost, of several
thousand dollars they built a flume, put in a hydraulic
plant, and gathered a pile of loose dirt to wash down the
flume, where the gold is gathered upon quicksilver. The
" sharks " raised $50,000 for a gold-dust fund. This dust was
run evenly over the quicksilver so that when the capitalists
returned, there was everything to show an enormously
rich placer-ground. The capitalists insisted upon a clean-up
after the first fortnight's run, which added so much more
joy to the sharks. This time the bait was swallowed whole,
string and all. The capitalists paid down promptly $250,000
for the ground. The sharks left the country. In a few
weeks nothing could be found but the amalgam of the
sharks.

An ingenious trick once baffled some experienced experts and came very near selling a mine. The mine was a
well developed one and had done great things in its day.
It was claimed that at the face of the tunnel, or where the
workings left off, there was still a fine showing of ore in
place to go on with. The experts found it as stated; on the
face or end of the tunnel there was a fine showing of ore,
and the probable amount in place and for the future was
duly measured up and estimated. It leaked out later that
this block of ore was only a thin screen purposely left, all
back of, and behind it, having been carefully worked out
and the opening for the miners ingress and egress skilfully
concealed. The mine was re-examined, the cheat discovered
and the reputation of the experts saved as well as many
thousands of dollars from the pockets of guileless investors.

This brief sketch of some of the ways of some miners, for
some regions and properties, would give an unfair idea of
some mines and miners as a whole, if it were supposed that
all miners are given to salting, and all properties for sale are
beset by a network of dishonest devices. On the contrary
many, very many, miners are as straight as a string and
hundreds of properties are to be examined without fear of
tampering. But it often happens that a miner, who in every
other relation of life, is as honest as the day, draws a line,
when it comes to the selling of a mine, which he considers
" fair game."

But, as elsewhere the world through, honesty pure and
simple is the right policy, and in the end would be found
the best paying one. For the notorious dishonesty connected with mines (much more common in the past than in
the present) scares away capitalists from investing, whilst

if truth and honesty were maintained, money would roll in freely.

One lesson at least may be learned from what we have said, and that is, that if in some cases a professional expert is ever taken in, what chances has a capitalist, ignorant of mines, to buy a mine on his own examination? What man ignorant of horseflesh would venture to buy a steed from a professional horse-jockey, without taking with him a friend who is knowing about horses?

How much more so in such a difficult and delicate problem as that of purchasing a mine, is it the duty of an investor never to purchase or induce his friends to purchase a mine, until he has employed the services of a competent expert to previously examine it. If the expert's fee should amount to a few hundreds, and after all he should decide on condemning the property, it is far better for the company to entail this expense, and perhaps lose this small sum, than to involve themselves in the loss of thousands of their own as well as other people's money in a bogus, worthless, or wildcat scheme.

CHAPTER XV.

PROSPECTORS' TOOLS AND HOW TO SHARPEN AND TEMPER THEM.

The principal tools a prospector takes into the field, are picks, drills, hammers and shovel.

A prospector, especially when climbing mountains, likes to be as light-handed and unencumbered as possible.

For his trip as a whole, he may carry several different tools packed on his donkey, but when he has arrived at a locality, the vicinity of which looks likely, he leaves most of his heavier tools in his temporary camp, or near to where he pickets his pack animal. He makes a short excursion up the mountain for a general reconnoitre, armed with nothing more than a light prospecting pick, weighing not more than three or four pounds. This little pick is about ten inches in length, with a handle about fifteen inches long; the longer portion is sharpened into a pick, and the shorter ends in a square faced hammer. We recommend a square sharp cornered face to the hammer, in preference to the bevelled face, as the sharp edges and corners are better adapted for breaking rock than the rounded or bevelled ends. This prospecting pick or geological pick and hammer, should be all of good steel, with a good sized eye to admit a springy handle of hickory. See Plate XCVII, Figs. 1, 1, 1.

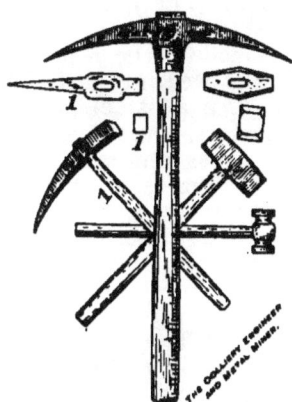

Armed with this little weapon he climbs the hillside, hunting for " float " or for rusty outcrops of ledges. Loose pieces of rock he cracks open with the hammer end, softer rock in place he explores with the pick. " When I am climbing over the hills," said an old weather-beaten prospector to me, " I want nothing but my little pick, then if I find anything likely ' in place,' I mark the spot, and go on, and at noon I come down to camp, or to where the ' burro ' is feeding, I take up my heavy digging pick and shovel and ' open up '; this will occupy me till evening at least, then if I find there is a ledge worth more thorough exploring, I leave my tools by the hole, and next morning bring up the drills, hammers and blasting outfit. But the first thing I would advise a tenderfoot, is to *get his eye trained*, trained to looking for float and observing mineral signs, trained to the whole business of close observation. Why! I myself, old hand as I am, after being away for some months about town or looking at other things, can't get my eye in and down to it for two or three days ; then it kind of comes natural.

Picks and Hammers.

PLATE XCVII.

" You must have an eye for float and rocks like an artist has an eye for color, and a musician, an ear for music. A tenderfoot had better go along with an old hand for a few days to get into training."

DESCRIPTION OF TOOLS, PICKS AND DRILLS.

Picks and drills are the main tools that need sharpening and tempering. The kind of sharpening and nature or degree of tempering depend upon the kind of work or kind

of rock to be worked, whether hard or soft, loose grained
or fine grained, siliceous or clayey. Drills, for example,
would have to be differently sharpened and tempered for
hard vitreous quartzite than for soft sandstone or hardened
clay. The same remark applies also to picks. Picks may
be double pointed or single, or with a hammer head called a
poll, if it is to be used for breaking rock. The main points
of.a pick are, strong cutting tips, stout eye and a tight
handle. The little prospecting pick is made of the best steel
throughout, but in the heavier pick, the wearing parts are
the tips, which should be replaceable. An all steel pick is
liable soon to be shortened up and useless, whilst the iron
pick eye, a 14 inch length of best iron, gives long service by
welding on tip ends, whenever desired. Professor Ihlseng,
in his " Manual of Mining," as also Mr. George Andre, in his
book on " Rock Blasting," give excellent descriptions of
tools used as well as the mode of sharpening and tempering
them; to them we are indebted for many of the details of
this article, and to their works we refer the reader for
further information on this subject. "The picks are
sharpened to form on an anvil, and commonly drawn to a
four sided pyramidal point, for hard rock, and a slim taper
for fissured rock, and a bluff taper to cut crisp ground, and
to a chisel end for chipping the ground. The eye is oval
and well surrounded with metal. All the strain of the pry-
ing falls on the eye, which must be true and stout."

DRILLS.

" The drill is a bar which has one cutter edge and one
hammer end. It is of round or octagonal steel. Drills may
be of various lengths, from a foot to four or five or even
more feet. For prospecting purposes two or three medium
short drills from two to four feet are generally enough, as
the prospector's business is rather to find than to develop.
In beginning to drill, it is common to use a short thick drill,
with a stout 'bull edge' rather than a thin, tapering one,
especially in hard rock; smaller sized, $i.\ e.$, narrower drills
may be used for increasing depth.

" The rock drill consists of chisel edge, bit, stock and
striking face. To allow the tool to free itself readily in the
bore hole, and to avoid introducing unnecessary weight
onto the stock, the bit is made wider than the latter. In
hard rock, the liability of the edge to fracture increases as
the difference of width ; the edge of the drill may be straight

or slightly curved, a straight edge cuts more freely than the curved ; a bull bit for hard rock is generally curved, a straight edge is weaker at the corners than the curved. The width of bits varies from 1 inch to 2½ inches. Figs. 1, 2, 1*a*, 2*b*, Plate XCVIII., show the straight and curved bits and angles of cutting edges for use in rock. The stock is octagonal in section. It is made in lengths varying from 20 inches to 42 inches. The shorter the stock, the more effectively it transmits the force of the blow. To insure the longer drills working freely in the hole, the width of the bit should be very slightly reduced in each length. Diameter of stock is less than the width of the bit generally by ⅜ of an inch.

" The smith cuts up the ' borer ' steel bars into desired lengths to form the bit, the end of the bar is heated and flattened out by hammering to a width a little greater than the diameter of the hole to be bored. The cutting edge is then hammered up with a light hammer to the requisite

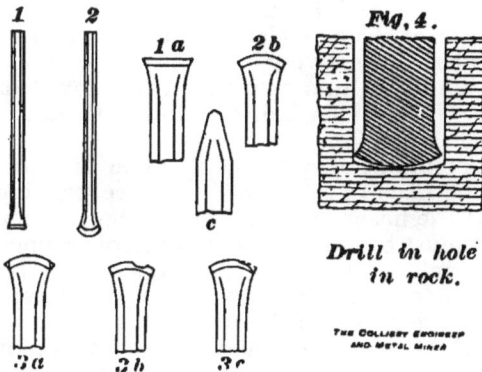

PLATE XCVIII.
Forms of Drills.

angle and corners beaten in to give the exact diameter of the bore hole intended. The drills are made in sets and the longer stocks will have a bit slightly narrower than the shorter ones for reasons already given. The edge is touched up with a file. Heavy hammering and high heats should be avoided. The steel should be well covered with coal, in making the heat, and protected from the raw air. Overheated or burned steel is liable to fly, and drills so injured are useless until the burned portion has been cut away. Care is required to form the cutting edge evenly, and of the full form. If the corners get hammered as in Fig. 3*a*, Plate XCVIII., they are said to be 'nipped' and the tool will not free itself in cutting. When a depression of the straight or curved line forming the edge occurs, as Fig. 3*b*, the bit is said to be 'backward' and when one of the corners is too far back, as Fig. 3*c*, it is spoken of as 'odd cornered.' Either of these defects causes the force of the blow to be thrown upon a portion only of the edge, which is thereby overstrained and liable to fracture."

SHARPENING TOOLS.

Professor Ihlseng in his " Manual of Mining" says : " The best fuel for blacksmithing may be a slightly caking coal, giving flame and high heat. Coke is hotter but harder to keep fire in. The fuel should be as free from sulphur as possible. White ash coal is better than red ash ; sulphur makes the iron hot short, and tends to produce scales. The coal should be clear of shale or slate, for they fuse and make a pasty cinder that is annoying."

A prospector away from civilization may have to use wood ; in that case he should use chips, and blow them with a portable bellows.

The prospectors who try to get along on as small an outfit as possible usually take one to three blasting powder cans and cut the heads out of all but the bottom one, and one head of that must be cut out ; these they place one on top of the other to make a furnace. They punch an inch and a half hole in the side of the bottom one at the bottom for draft and to put in the points of the tools to heat them.

They use charcoal for fuel and then a chunk of steel or railroad iron about 6 inches long serves for an anvil. Some take a small bellows and anvil with them. For tempering drills they give the drill, when red, a plunge in water. After two or three rubs on wood, to brighten it, they hold it up to the light and watch it until it takes on a straw color. Then they dip it in water again. For picks a blue color is the most satisfactory in general.

" Steel is a compound of iron and carbon and its homogeneity and presence of carbon imparts to it a capability of hardening and tempering to a degree depending on the temperature of the heating and subsequent cooling. As the amount of carbon increases, the melting point of the iron decreases, and this greater fusibility reduces its welding quality.

" A steel is called ' hardened ' when it has been suddenly cooled and thereby become as hard as possible. This is owing to the presence of carbon, for pure malleable iron is not affected by the operation, while both steel and cast iron are to a marked degree.

" The operation consists in forging the steel to a certain degree of temperature, and then plunging it into some fluid which abstracts the heat from the tool. The quicker it is done, and the greater the difference of temperature, the harder is the tool. Either water or oil is used ; both volatilize at a temperature much below that of the immersed tools, so the hardening takes place in a vapor ; oil generally produces the best effects. On the first plunge the metal is chilled and coated with soot, after which, a slow process of cooling takes place."

TEMPERING.

" Tempering follows hardening, whereby the steel is sub-jected to a subsequent lower heat, which softens it, and removes its brittleness. When the hardened iron is slowly reheated, its surface gradually assumes phases of color, beginning with a light straw, passing through shades of yellow, brown, purple, blue and red. At a cherry-red heat, the original color before hardening, the effects of the chilling are practically removed.

" Tempering consists in carrying the second heat to one of the above mentioned colors, according to the amount of the brittleness to be annealed. This depends upon the use to which the article is to be put. A second stage of the opera-tion finishes the job. The aforementioned reheat, goes on a little way beyond the desired color. The tool is carefully plunged part way into the water or oil, till the disappearance of the steam indicates that it is cold, when another portion of the distance is further immersed for a moment. The tool is withdrawn, the scales rubbed off and the heat of the remaining portion draws to the edge, until it has assumed the proper tempering color. It is then thoroughly cooled. The idea that the steel is cooler at a blue, than at a yellow, in final drawing, is erroneous ; for more of the heat is conducted from the red portion to the point, than it radiates to the air, and the first heat to the edge only gives a yellow ; with more, it becomes purple, and so on. Hardened drill and pick points are treated in in this way, 4" of the end being heated to a yellow ; and, in thirds. the tempering is proceeded with as above.

"Care should be taken that the plunged tool while tempering, be not held too long a time at a certain color line, as it
has a tendency to break at that point. The tool should be
slightly waved in the water. 'Pieces' which are to be tempered throughout must be allowed to soak, *i. e.*, become
uniformly hot, before plunging.

"The proper color for a given ground, is only ascertained
by experience. Generally speaking, the picks and drills are
stopped at a 'straw,' if intended for hard ground ; at a blue,
for mild ground. The toughness of the steel should be preserved as much as possible, therefore select the lowest color
compatible with the service to be performed. A high carbon
steel is given a lighter color than steel of low carbon.

"A pick is made of a square iron bar 14″ × 1¼″ heated at
the middle, and then struck endwise, till about 1½″ across.
This spot is softened and at red heat, cut open, and swelled
by a drift to form the eye. This is then slit at the ends, and
softened, while a 6″ length of pick steel is being heated.
When ready, this steel is tongued into the iron, and hammered. A reheating with borax, and a hammering complete
the weld, after which the picks are sharpened and tempered,
no signs of the weld should be visible."

"Pick-steel" is a special steel that can be had in bars 1¼″
or 1½″ × ⅝″ or ¾″ and used only for tips.

Steel bars for drills come in lengths of about 14 feet each
and from ⅝″ to 2″ diameter. The American "Black Diamond" brand is a favorite. The bars are cut into pieces as
long as can conveniently be used, *e. g.* 30″ and 36″. The bits
are wider than the tool, to prevent it sticking to the hole.
They are widened according to pattern, so they can "follow"
well. The first drill has the widest bit; the followers narrower ones. In hard rock the flare is smaller than that in
soft rock.

"The temper is a lighter color for hard than for soft rock.
If the edges of the returned drills are cracked or broken the
steel is too brittle, and should be made softer or other coal
used. If the edges blunt much by wearing round, they are
all right, though a harder temperature may give them longer
life. Cast steel borers are never heated above a cherry.
They are annealed at the striking end."

PRACTICAL SUGGESTIONS AND POINTS BY A BLACKSMITH.

A prospector must have something to act as an anvil, a
hard pebble wont do, he can carry a small anvil or a chunk

of railroad iron. A small hand bellows or even a portable forge worked with a crank will make his outfit complete. The following practical hints I picked up from a blacksmith whilst watching him at work tempering both picks and drills for some prospectors. He said: "You must temper your drill according to the character of the rocks.

"For hard rock, use a short thick edged 'bull bit' which will stand a high brittle temper such as 'straw.' For picks, a light blue color is a good temper, rather than 'straw' which is too brittle. Cherry red is the heat of your bar, not hotter; laying this on the anvil and hammering it well all over gives it toughness. If blisters show on the steel you must hammer it over again. By occasionally dipping your hammer in water and then striking with it, you get the steel down to a fine grain. When you are dipping for tempering, put the point in the water, that cools the point, and the heat runs the color down to the cool point; when the color reaches the tint you want, then is your time to cool off quickly. The color progresses from a white or pale straw to copper color, to blue. Copper tint is a good one to stop at for a drill,— blue, for a pick. The right moment to stop and cool is just at the turning point from one color to another."

He took a piece of steel, heated it to cherry red, laid it on the anvil and pounded it lightly with his hammer all over, to toughen it by blows, occasionally dipping his hammer in the water to "water temper" it; this further toughens it, by partially cooling it. Now the bar was again put in the fire and heated to a cherry red, care being taken not to keep the bar too long in the fire, as that would tend to take its toughness out, or produce blisters. The bar was plunged about an inch into the water, and then rubbed against a brick, to show the colors plainer. These passed from the point upwards, gradually through the colors we have mentioned; to arrest it by suddenly cooling off at "straw," would make it too brittle for ordinary drills, except a "bull drill." Now the "straw" turns into a copper hue, a good point to cool off for a *drill*. Now it passes into a blue, at this point it would be well to cool off for a *pick*. The edge of a drill is almost of secondary importance to the *sharpness of the projecting corners;* when these are gone, the drill is used up, and clogs in the hole. Some rocks like sandstone will, by reason of the quartz in them, wear off the corners very rapidly, others, like limestone or granite, less rapidly.

Another blacksmith advised me not to dip (as is commonly done) the point *only an inch* in water as it is apt in use to break at the water line, but plunge it *all over* in the water. " Who shall decide when doctors disagree ?" A prospector should take with him a regular blacksmith's hammer for sharpening, as well as the 4 or 5-lb. hammer he uses for striking drill or the rock.

CHAPTER XVI.

SOME ELEMENTS OF MINING LAW RELATING TO PROSPECTING.

A prospector would do well to acquaint himself with a few elements of mining law, so we will give a few samples of Colorado mining law for his benefit.

Extent of Lode or Claim.—The length of any lode may equal, but not exceed, 1,500 feet along the vein.

Dimensions.—The width of lode claims in Gilpin, Clear Creek, Boulder and Summit counties, shall be 75 feet on each side of the center of the vein or crevice.

Certificate of Location.—The discoverer of a lode shall, within three months from the date of discovery, record his claim in the office of the recorder of the county in which such lode is situated, by a location certificate, which shall contain :

1st. The name of the lode.
2d. The name of the locator.
3d. The date of the location.
4th. The number of feet in length claimed on each side of the center of the discovery shaft.
5th. The general course of the lode as near as may be.

Discovery Shaft.—Before filing such location certificate, the discoverer shall locate his claim by first sinking a discovery shaft on the lode, to the depth of at least 10 feet, or deeper if necessary, to show a well-defined crevice.

Second, by posting at the point of discovery on the surface, a plain sign or notice containing the name of the lode, the name of the locator and the date of the discovery.

Third, by marking the surface boundary line of the claim.

Staking.—Such surface boundaries shall be marked by six substantial posts, hewed or marked on the side or sides of which are in toward the claim, and sunk in the ground, to wit, one at each corner, and one at the center of each side line. Where it is impossible on account of bed rock, or precipitous ground, to sink such posts, they may be placed in a pile of stones.

Open Cuts.—Any open cut or cross-cut tunnel, or tunnel which shall cut a lode at the depth of ten feet below the surface, shall hold it, the same as if a discovery shaft were sunk thereon, or an adit of at least ten feet along the lode from the point where the lode may be in any manner discovered, shall be equivalent to a discovery shaft.

Time.—The discoverer shall have 60 days from the time of uncovering or disclosing a lode, to sink a discovery shaft thereon.

Construction of Certificate.—The location certificate of any lode claim shall be constructed to include all surface ground within the surface lines thereof, and all lodes and ledges throughout their entire depth, the top or "apex" of which lies inside of such lines extending downward vertically, with such parts of all lodes or ledges as continue to dip beyond the side-lines of the plane, but shall not include any portion of such lodes or ledges beyond the end lines of the claim, or at the end-lines continued, whether by dip or otherwise, or beyond the side-lines in any other manner than by the dip of the lode.

Cannot be Followed.—If the top or "apex" of a lode in its longitudinal course extends beyond the exterior lines of the claim at any point on the surface, or as extended vertically downward, such lode may not be followed in its longitudinal course beyond the point where it is intersected by the exterior lines.

Proof of Development.—The amount of work done, or improvements made during each year shall be that prescribed by laws of the United States.

Placer Mining Claims.—The discoverer of a placer claim shall, within 30 days from the date of discovery, record his claim in the office of the recorder of the county in which said claim is situated, by a location certificate, which shall contain :

1st. The name of the claim, designating it as a placer claim.

2d. The name of the locator.

3d. The date of the location.

4th. The number of feet or acres claimed.

5th. The description of the claim by such reference to natural objects or permanent monuments as shall identify the claim.

Before filing such location certificate, the discover shall locate his claim :

1st. By posting upon such claim a plain sign or notice containing the name of the claim and of the locator, the date of discovery, and number of acres or feet claimed.

2d. By marking the surface boundaries with substantial posts sunk in the ground, one at each angle of the claim.

On each placer claim of 160 acres, not less than 100 dollars' worth of labor shall be done by the first of August each year, and upon less or more ground a sum in proportion.

INDEX.

LIST OF AUTHORS AND WORKS REFERRED TO.

Balch, W. R. } Mines, Miners and Mining Interests of the United
Balch, A. } States.
Cross, Whitman.—Geology of Cripple Creek.
Dana, J. D.—Geology and Mineralogy.
Emmons, S. F.—Geology and Mining Industry of Leadville.
Farish, J. B.—A Typical Boulder County Mine.
Geikie, A.—Hand Book of Field Geology.
Guiterman, F.—Red Cliff Gold Deposits.
Ihlseng, M. C.—Manual of Mining.
Judd, J. W.—Volcanoes.
Kemp, J. F.—Ore Deposits.
Lakes, A.—Geology of Colorado and Western Ore Deposits.
Le Conte, J.—Geology.
Lock, A. G.—Gold, its Occurrence and Extraction.
Lock, C. G. W.—Practical Gold Mining.
Penrose, Arthur.—Ore Deposits of Cripple Creek.
Phillips, J. A.—Ore Deposits.
Williams, Albert, Jr.—Mineral Resources of the United States.